Guideline for Salinity Assessment, Mitigation and Adaptation Using Nuclear and Related Techniques

Mohammad Zaman • Shabbir A. Shahid
Lee Heng

Guideline for Salinity Assessment, Mitigation and Adaptation Using Nuclear and Related Techniques

Mohammad Zaman
Soil and Water Management & Crop
Nutrition Section, Joint FAO/IAEA
Division of Nuclear Techniques in
Food and Agriculture, Department
of Nuclear Sciences & Applications
International Atomic Energy Agency
(IAEA)
Vienna, Austria

Shabbir A. Shahid
Senior Salinity Management Expert, Freelancer
Dubai, UAE

Lee Heng
Soil and Water Management & Crop
Nutrition Section, Joint FAO/IAEA
Division of Nuclear Techniques in
Food and Agriculture, Department
of Nuclear Sciences & Applications
International Atomic Energy Agency
(IAEA)
Vienna, Austria

https://doi.org/10.1007/978-3-319-96190-3

Salt-tolerant grass growing under saline condition

Foreword

Soil salinity is a major global issue owing to its adverse impact on agricultural productivity and sustainability. Salinity problems occur under all climatic conditions and can result from both natural and human-induced actions. Generally speaking, saline soils occur in arid and semi-arid regions where rainfall is insufficient to meet the water requirements of the crops, and leach mineral salts out of the root-zone. The association between humans and salinity has existed for centuries and historical records show that many civilizations have failed due to increases in the salinity of agricultural fields, the most known example being Mesopotamia (now Iraq). Soil salinity undermines the resource base by decreasing soil quality and can occur due to natural causes or from misuse and mismanagement to an extent which jeopardizes the integrity of soil's self-regulatory capacity.

Soil salinity is dynamic and spreading globally in over 100 countries; no continent is completely free from salinity. Soil salinization is projected to increase in future climate change scenarios due to sea level rise and impact on coastal areas, and the rise in temperature that will inevitably lead to increase evaporation and further salinization. There is a long list of countries where salt-induced land degradation occurs. Some well-known regions where salinization is extensively reported include the Aral Sea Basin (Amu-Darya and Syr-Darya River Basins) in Central Asia, the Indo-Gangetic Basin in India, the Indus Basin in Pakistan, the Yellow River Basin in China, the Euphrates Basin in Syria and Iraq, the Murray-Darling Basin in Australia, and the San Joaquin Valley in the United States.

The objective of this guideline is to develop protocols for salinity and sodicity assessment, and the role of isotopic nuclear and related techniques to develop mitigation and adaptation measures to use saline and sodic soils sustainably. We have focused on important issues related to salinity and sodicity and have described these in an easy and user friendly way. The information has been compiled from latest published literature and from authors' publications specific to the subject matter. This guideline is an outcome of a joint publication between the Soil and Water Management & Crop Nutrition Section, Joint FAO/IAEA Division of Nuclear Techniques in Food and Agriculture, International Atomic Energy Agency (IAEA),

Vienna, Austria, and a freelance senior salinity management expert based in the United Arab Emirates.

We hope that this guideline will be an excellent contribution to the science and enhance the knowledge of those seeking information to assess and diagnose salinity problem at the landscape and farm levels and the role of nuclear and isotopic techniques in developing strategies to use these marginal soils sustainably.

Soil and Water Management & Crop Nutrition Section, Joint FAO/IAEA Division of Nuclear Techniques in Food and Agriculture
Department of Nuclear Sciences & Applications
International Atomic Energy Agency (IAEA)
Vienna, Austria

Mohammad Zaman

Senior Salinity Management Expert, Freelancer, Dubai, UAE

Shabbir A. Shahid

Soil and Water Management & Crop Nutrition Section, Joint FAO/IAEA Division of Nuclear Techniques in Food and Agriculture
Department of Nuclear Sciences & Applications
International Atomic Energy Agency (IAEA)
Vienna, Austria

Lee Heng

Acknowledgements

We are thankful to Prof. Pharis, R.P., Department of Biological Sciences, University of Calgary, Canada, and Dr. Shazia Zaman, University of Canterbury, for their critical review, feedback and editorial comments in the preparation of this book. We also thank Ms. Marlies Zaczek of the Soil and Water Management & Crop Nutrition Section, Joint FAO/IAEA Division, International Atomic Energy Agency (IAEA), Vienna, Austria, for her help in formatting this document and Ms. Petra Nabil Salame, PMO of Asia and the Pacific Section 2, Division for Asia and the Pacific Department of Technical Cooperation for her financial support.

Contents

About the Authors

Mohammad Zaman is working as Soil Scientist/Plant Nutritionist at the Soil and Water Management & Crop Nutrition (SWMCN) Section, Joint FAO/IAEA Division of Nuclear Techniques in Food and Agriculture, International Atomic Energy Agency (IAEA), Vienna, Austria. Zaman's work is covering countries in Africa, Asia, Middle East, Europe, and Central and Latin America on developing climate smart agricultural practices for enhancing food production and environmental sustainability. Prior to joining the SWMCN Section, he worked for 19 years in integrated plant nutrient management of irrigated and dryland farms at different research, academic, commercial, and international organizations in both developing and developed countries.

Zaman completed a Ph.D. degree at Lincoln University, Canterbury, New Zealand, on soil nitrogen (N) mineralization and its relationship to soil microbial and enzyme activities in grasslands under different management practices. After completing his Ph.D., Zaman continued post-doctoral research at Lincoln University and Chiba University, Japan, in the areas of soil fertility, soil water quality, and greenhouse gases, using both conventional and stable isotopic techniques. After 2 years in Japan, Zaman went back to New Zealand to take up a position as a researcher in the National Institute of Water & Atmospheric Research (NIWA), Hamilton. Later, he moved to the farmers owned Cooperative Fertilizer Industry, firstly as a Senior Scientist and then as a Research Manager to increase the adaptive capabilities of soil/plant systems to climate change and to enhance nutrient use efficiency on farms. This research included mitigating greenhouse gas emissions in an integrated plant animal system, developing new decision support systems and tools, identifying novel fertilizer products to increase farm productivity and resource use efficiency, and minimizing nutrient losses to waterways and the atmosphere. Zaman is author and coauthor of over 65 research publications in refereed journals.

Shabbir Ahmad Shahid is Senior Salinity Management expert a freelancer based in the United Arab Emirates. He had served in different capacities in a number of organizations, including Senior Salinity Management Scientist in ICBA, Dubai, UAE, Associate Professor of Soil Science at the University of Agriculture, Faisalabad, Pakistan, Associate Research Scientist at the Kuwait Institute for Scientific Research, Kuwait, Manager of Soil Resources Department at the Environment Agency, Abu Dhabi, UAE. He has more than 36 years of experience in applied agricultural research in many countries and regions. Dr. Shabbir holds a Ph.D. in "soil micromorphology of salt-affected soils" from Bangor University, Wales, UK. He is author and coauthor of over 160 research publications, including five edited books published by Springer, principal author of two books and two manuals, 52 peer-reviewed journal papers, 31 book chapters, and 27 conference proceedings. He also authored and coauthored over 30 scientific reports. He is member of a number of scientific committees, international advisory boards of international conferences, and editorial boards of scientific journals. He is life member of World Association of Soil and Water Conservation (WASWAC). He is a recipient of Sir William Roberts and David A Jenkins awards. Shabbir is pioneer in introducing soil survey research in the UAE. He with his associates discovered anhydrite soil in UAE and formally added to 12 edition of US Keys to Soil Taxonomy. He also authored UAE Keys to Soil Taxonomy and chaired the Emirates Soil Museum Committee to establish a unique soil museum at ICBA. His research priorities include: soil surveys and salinity mapping, reclamation of salt-affected soils, integrated soil fertility management, agricultural intensification through soil health improvement using organic and inorganic amendments, conservation agriculture and climate smart agriculture, environmental impact assessment, land degradation, and carbon sequestration.

Lee Heng has a Ph.D. in soil science from Massey University, Palmerston North, New Zealand, and has more than 25 years' experience in soil-plant-water interactions, agricultural water management and water use efficiency, integrated nutrient-water interactions, and diffuse pollution control for sustainable agricultural production systems, at both national and international levels. Her work is covering countries in Africa, Asia, Europe, and Central and Latin America on sustainable land and water management for climate smart agriculture and the efficient use and conservation of agricultural resources for enhancing food production and environmental sustainability. For the past 18 years, Heng works at the Soil and Water Management & Crop Nutrition (SWMCN) Subprogram in the Joint FAO/IAEA Division of Nuclear Techniques in Food and Agriculture. Currently, she is the Head of SWMCN Section, which assists scientists in Member States in the development, validation and dissemination of a range of soil, water, and crop management technology packages through the use of nuclear and nuclear-related techniques.

Prior to her current assignment with the Joint FAO/IAEA Division, Heng worked as a research associate in the Department of Agriculture and Forestry at the University of Melbourne, Australia. Her work involved measurement and modeling the transport of reactive solutes in soils, conducting field study on soil water and nitrogen dynamics under temperate pastures. She also worked with Landcare Research in New Zealand. Heng has won several awards for her professional achievements including the USDA-Agricultural Research Service (ARS) Outstanding Sustained Effort Technology Transfer Award in 2012 and several IAEA Merit Awards in different categories. Heng is author and coauthor of over 60 research publications in refereed journals.

Acronyms and Abbreviations

BNF	Biological Nitrogen Fixation
CA	Conservation Agriculture
CEC	Cation Exchange Capacity
CID	Carbon Isotope Discrimination
CRDS	Cavity Ring-Down Spectroscopy
CSA	Climate Smart Agriculture
DEM	Digital Elevation Model
dS/m	deci Siemens per meter
E	Evaporation
EAD	Environment Agency Abu Dhabi
EC	Electrical Conductivity
ECa	Apparent Electrical Conductivity
ECe	Electrical Conductivity of Soil Saturation Extract
ECiw	Electrical Conductivity of Irrigation Water
EDXRA	Energy Dispersive X Ray Analyses
EMI	Electromagnetic Induction
ESP	Exchangeable Sodium Percentage
ET	Evapotranspiration
FAO	Food and Agriculture Organization of the United Nations
fc	Field Capacity
GCC	Gulf Cooperation Council
GIS	Geographic Information System
GR	Gypsum Requirement
IAEA	International Atomic Energy Agency
ICBA	International Center for Biosaline Agriculture
IDW	Inverse Distance Weighted
IM	Picarro Induction Module
IRMS	Isotope Ratio Mass Spectrometry
ISFM	Integrated Soil Fertility Management
ISRP	Integrated Soil Reclamation Program

LR	Leaching Requirement
mmhos/cm	millio mhos per centimeter
mS/cm	milli Siemens per centimeter
NRM	Natural Resources Management
OPUS	Options for the Productive Use of Salinity
ppm	parts per million
RS	Remote Sensing
RSC	Residual Sodium Carbonates
RTASLS	Realtime Automated Salinity Logging System
S	Siemens
SAR	Sodium Adsorption Ratio
SBCS	Serial Biological Concentration of Salts
SDI	Subsurface Drip Irrigation
SI	Standard International
SWMCN	Soil and Water Management and Crop Nutrition
TDS	Total Dissolved Solutes
TM	Thematic Mapper
TSS	Total Soluble Salts
μS/cm	micro Siemens per centimeter
USSL	United States Salinity Laboratory
WDXRA	Wavelength Dispersive X Ray Analyses
WUE	Water Use Efficiency

List of Figures

List of Plates

List of Tables

Chapter 1
Introduction to Soil Salinity, Sodicity and Diagnostics Techniques

Shabbir A. Shahid, Mohammad Zaman, and Lee Heng

Abstract It is widely recognized that soil salinity has increased over time. It is also triggered with the impact of climate change. For sustainable management of soil salinity, it is essential to diagnose it properly prior to take proper intervention measures. In this chapter soil salinity (dryland and secondary) and sodicity concepts have been introduced to make it easier for readers. A hypothetical soil salinity development cycle has been presented. Causes of soil salinization and its damages, socio-economic and environmental impacts, and visual indicators of soil salinization and sodicity have been reported. A new relationship between ECe (mS/cm) and total soluble salts (meq/l) established on UAE soils has been reported which is different to that established by US Salinity Laboratory Staff in the year 1954, suggesting the latter is specific to US soils, therefore, other countries should establish similar relationships based on their local conditions. Procedures for field assessment of soil salinity and sodicity are described and factors to convert EC of different soil: water (1:1, 1:2.5 & 1:5) suspensions to ECe from different regions are tabulated and hence providing useful information to those adopting such procedures. Diversified salinity assessment, mapping and monitoring methods, such as conventional (field and laboratory) and modern (electromagnetic-EM38, optical-thin section and electron microscopy, geostatistics-kriging, remote sensing and GIS, automatic dynamics salinity logging system) have been used and results are reported providing comprehensive information for selection of suitable methods by potential users. Globally accepted soil salinity classification systems such as US Salinity Lab Staff and FAO-UNESCO have been included.

Keywords Salinity · Sodicity · Diagnostics · Electromagnetic · Geostatistics · GIS · Kriging · Electron microscopy

1 Introduction

Soil is a non-renewable resource; once lost, can't be recovered in a human lifespan. Soil salinity, the second major cause of land degradation after soil erosion, has been a cause of decline in agricultural societies for 10,000 years. Globally about 2000 ha

M. Zaman et al., *Guideline for Salinity Assessment, Mitigation and Adaptation Using Nuclear and Related Techniques*, https://doi.org/10.1007/978-3-319-96190-3_1

of arable land is lost to production every day due to salinization. Salinization can cause yield decreases of 10–25% for many crops and may prevent cropping altogether when it is severe and lead to desertification. Addressing soil salinization through improved soil, water and crop management practices is important for achieving food security and to avoid desertification.

1.1 *What Is Soil Salinity?*

Soil salinity is a measure of the concentration of all the soluble salts in soil water, and is usually expressed as electrical conductivity (EC). The major soluble mineral salts are the cations: sodium (Na^+), calcium (Ca^{2+}), magnesium (Mg^{2+}), potassium (K^+) and the anions: chloride (Cl^-), sulfate (SO_4^{2-}), bicarbonate (HCO_3^-), carbonate (CO_3^{2-}), and nitrate (NO_3^-). Hyper-saline soil water may also contain boron (B), selenium (Se), strontium (Sr), lithium (Li), silica (Si), rubidium (Rb), fluorine (F), molybdenum (Mo), manganese (Mn), barium (Ba), and aluminum (Al), some of which can be toxic to plants and animals (Tanji 1990).

From the point of view of defining saline soils, when the electrical conductivity of a soil extract from a saturated paste (ECe) equals, or exceeds 4 deci Siemens per meter (dS m^{-1}) at 25 °C, the soil is said to be saline (USSL Staff 1954), and this definition remains in the latest glossary of soil science in the USA.

1.1.1 Units of Soil Salinity

Salinity is generally expressed as total dissolved solutes (TDS) in milli gram per liter (mg l^{-1}) or parts per million (ppm). It can also be expressed as total soluble salts (TSS) in milli equivalents per liter (meq l^{-1}).

The salinity (EC) was originally measured as milli mhos per cm (mmho cm^{-1}), an old unit which is now obsolete. Soil Science has now adopted the Systeme International d'Unites (known as SI units) in which mho has been replaced by Siemens (S). Currently used SI units for EC are:

- milli Siemens per centimeter (mS cm^{-1}) or
- deci Siemens per meter (dS m^{-1})

The units can be presented as:

1 mmho cm^{-1} = 1 dS m^{-1} = 1 mS cm^{-1} = 1000 micro Siemens per cm (1000 μS cm^{-1})

- EC readings are usually taken and reported at a standard temperature of 25 °C.
- For accurate results, EC meter should be checked with 0.01 N solution of KCl, which should give a reading of 1.413 dS m^{-1} at 25 °C.

No fixed relationship exists between TDS and EC, although a factor of 640 is commonly used to convert EC (dS m^{-1}) to approximate TDS. For highly concentrated solutions, a factor of 800 is used to account for the suppressed ionization effect on EC.

Similarly, no one relationship exists between ECe and total soluble salts (TSS), although a factor of 10 is used to convert ECe (dS m^{-1}) to TSS (expressed in meq l^{-1}) in the EC range of 0.1–5 dS m^{-1} (USSL Staff 1954). One relationship between ECe and TSS is presented in the Agriculture Handbook 60 (USSL Staff 1954). This relationship was developed using USA soils and has been widely used (worldwide) for over six decades. No efforts have been made to validate this relationship in other soils, though recently Shahid et al. (2013) have published a similar relationship for sandy desert soils ranging from low salinity (desert sand) to hyper-saline soils (coastal lands) in the Abu Dhabi Emirate. This latter work established a relationship between ECe and TSS which differs significantly from that of USSL Staff (1954), thus, opening the way for other countries to develop country-specific relationships which will allow better prediction and management of their saline and saline-sodic soils.

1.1.2 Why Total Soluble Salts *Versus* ECe Relationship Is Required?

Laboratories in some developing countries do not generally have modern equipment, i.e. flame emission spectrophotometer (FES), atomic absorption spectrophotometer (AAS), or inductively coupled plasma (ICP) in order to analyze soil saturation extracts or water samples for soluble Na^+ to determine sodicity (sodium adsorption ratio – SAR). In contrast, Ca^{2+} and Mg^{2+} are easy to measure using a titration method, one which does not require modern instruments. Currently, these laboratories in many developing countries determine soluble Na^+ by calculating the difference between the total soluble salts (TSS) and the quantities of Ca^{2+} + Mg^{2+} in order to make the analyses affordable, as below:

$$\mathbf{Na^+ = [(\text{Total soulble salts})\ \text{minus}\ (Ca^{2+} + Mg^{2+})]}$$

The TSS are recorded from a graph [see Fig. 4, page 12 of the Agriculture Handbook 60 (USSL Staff 1954)] by using the ECe value (Fig. 1.1). The Na^+ amount is then used to determine SAR so that exchangeable sodium percentage (ESP) can be calculated as:

$$\mathrm{SAR} = \frac{\mathrm{Na^+}}{\sqrt{\frac{1}{2}\left(\mathrm{Ca^{2+}} + \mathrm{Mg^{2+}}\right)}}$$

$$\mathrm{ESP} = \frac{[100\,(-0.0126 + 0.01475 \times \mathrm{SAR})]}{[1 + (-0.0126 + 0.01475 \times \mathrm{SAR})]}$$

Where, each of Na^+, Ca^{2+} + Mg^{2+} concentrations are expressed in milli equivalents per liter (meq l^{-1}) and SAR is expressed as (milli moles per liter)$^{0.5}$ (mmoles l^{-1})$^{0.5}$.

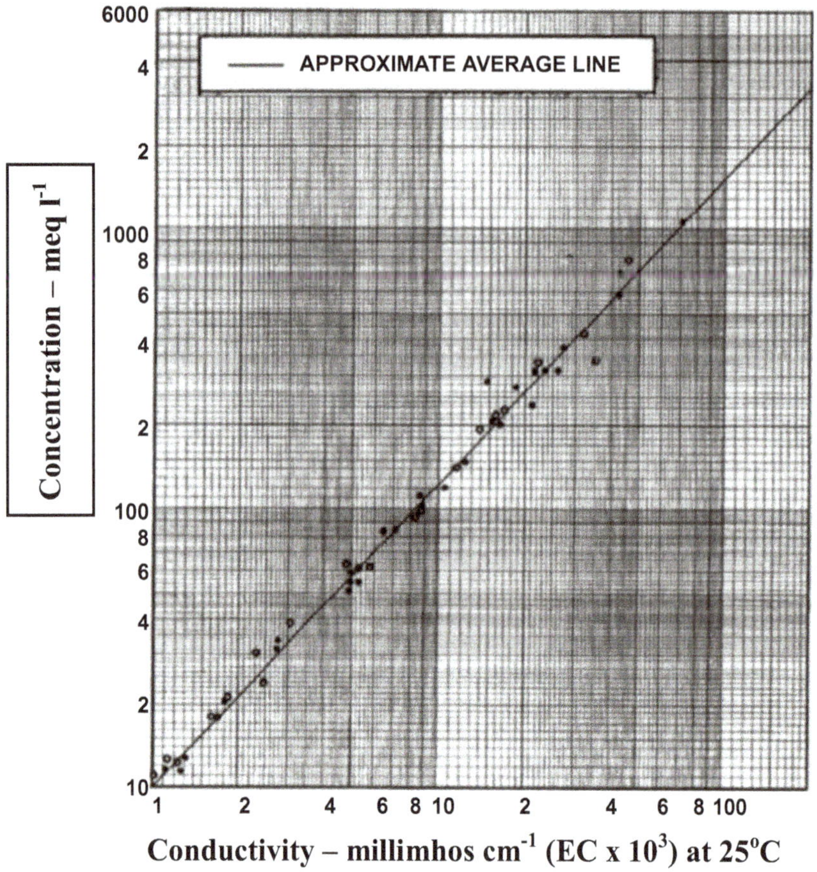

Fig. 1.1 Relationship between total soluble salts (TSS) on y-axis and ECe on x-axis. (Source: Fig. 4, page 12, Agriculture Handbook 60 (USSL Staff 1954))

In the above method of determining Na^+ by calculating the difference between TSS and Ca^{2+} + Mg^{2+}, any K^+ amounts present are added to the Na^+ (which is thus overestimated). It should be noted that the TSS versus ECe curve developed by USSL Staff (1954) was developed for the Western North American soils and, thus, may or may not be representative of soils of other countries. Hence, using such a practice may lead one to overestimate the sodicity hazard in irrigation waters or in saturation extracts of soils. This could lead to incorrect predictions and the use of inappropriate management options.

The finding of Shahid et al. (2013) has revealed an appreciable difference between the straight line (TSS versus ECe) determined from USSL Staff (1954) relative to protocols established by Shahid et al. (2013), shown in Figs. 1.1 and 1.2.

The TSS/ECe ratio, thus, ranges between 10 (at ECe 1 dS m^{-1}) and 16 (at ECe 200 dS m^{-1}) based on USSL Staff (1954) relationship (Fig. 1.1). In contrast, use of the relationship obtained from the methods developed by Shahid et al. (2013), the TSS/ECe ratio ranged between 10 (at 1 dS m^{-1}), 11.38 (at ECe 200 dS m^{-1}) and 12 (at ECe 500 dS m^{-1}). A comparative representation is shown in Figs. 1.3 and 1.4.

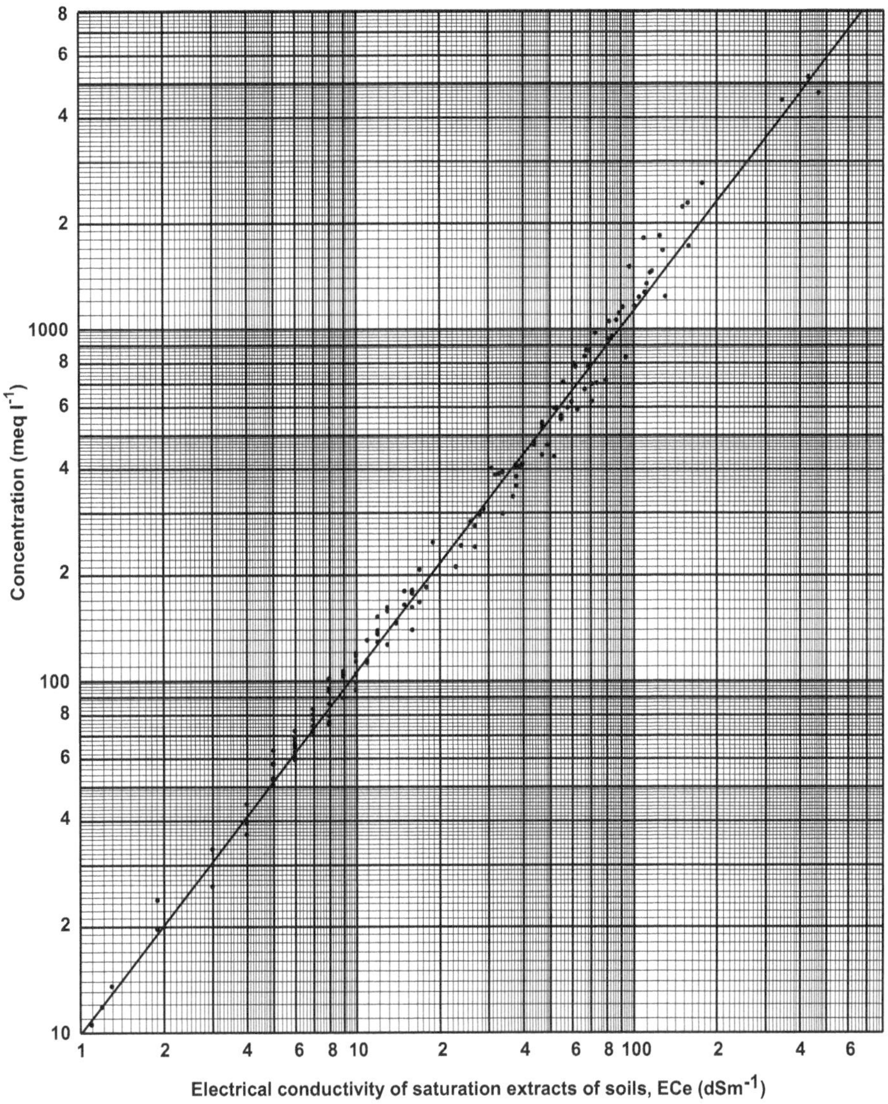

Fig. 1.2 Relationship between TSS and ECe (from Shahid et al. 2013)

In order to test the above lines to determine soil sodicity, Shahid et al. (2013) gave three examples using one soil type, as below.

Example 1

Determination of sodium adsorption ratio (SAR) accomplished by analyzing soil saturation extract for ECe (using an EC meter), and soluble Na^+, Ca^{2+}, Mg^{2+} determined by using an atomic absorption spectrophotometer.

ECe	$= 51$ dS m^{-1}
Soluble Na^+	$= 480$ meq l^{-1}
Ca^{2+}	$= 50$ meq l^{-1}
Mg^{2+}	$= 38$ meq l^{-1}
SAR	$= 72.4$ (mmoles l^{-1})$^{0.5}$

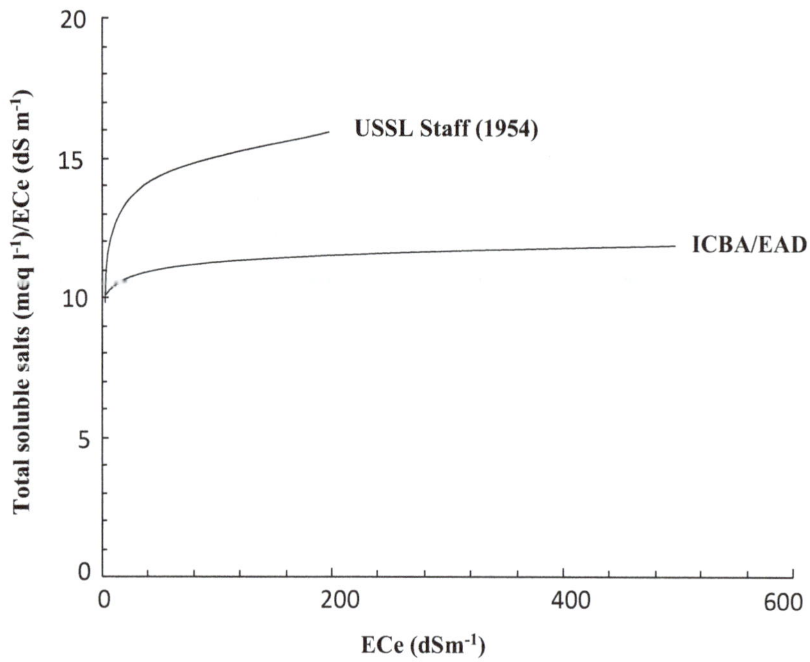

Fig. 1.3 A comparison of the relationship between total soluble salts (TSS)/ECe established using the Agriculture Handbook 60 curve (USSL Staff 1954) for the soils of Abu Dhabi Emirate and the relationship established for the same soils by the ICBA/EAD curve (Shahid et al. 2013)

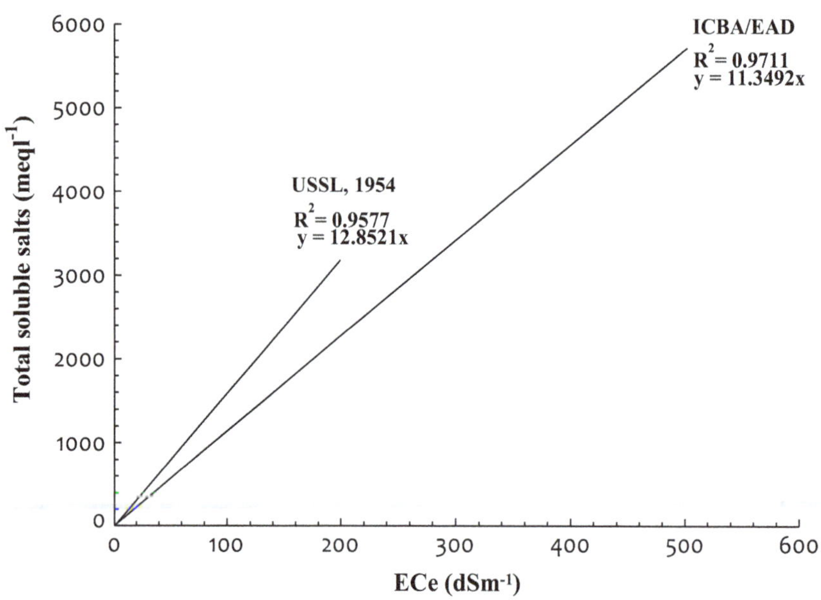

Fig. 1.4 Average lines showing the relationship between ECe and total soluble salts (TSS) from the USSL method (line from Fig. 1.1, above) for the soils of Abu Dhabi Emirate using the method developed by Shahid et al. (2013) using the average line adapted from Fig. 1.2, above

Example 2

Determination of sodium adsorption ratio (SAR) accomplished by analyzing soil saturation extract for ECe (using an EC meter), soluble Ca^{2+} and Mg^{2+} (by titration procedure) and soluble Na^+ estimated by calculating the difference between TSS and $Ca^{2+} + Mg^{2+}$ using USSL Staff (1954) relationship (Fig. 1.1).

ECe	$= 51$ dS m^{-1}
TSS	$= 720$ meq l^{-1} (from Fig. 1.1)
Soluble Na^+	$= 632$ meq l^{-1} (by difference, i.e. 720–88 = 632)
Ca^{2+}	$= 50$ meq l^{-1}
Mg^{2+}	$= 38$ meq l^{-1}
SAR	$= 95.32$ (mmoles $l^{-1})^{0.5}$

Example 3

Determination of sodium adsorption ratio (SAR) accomplished by analyzing soil saturation extract for ECe (using an EC meter), soluble Ca^{2+} and Mg^{2+} (by titration procedure) and soluble Na^+ by calculating the difference using the relationship developed by Shahid et al. (2013) (Fig. 1.2).

ECe	$= 51$ dS m^{-1}
TSS	$= 560$ meq l^{-1} (from Fig. 1.2)
Soluble Na^+	$= 472$ meq l^{-1} (by difference, i.e. 560–88 = 472)
Ca^{2+}	$= 50$ meq l^{-1}
Mg^{2+}	$= 38$ meq l^{-1}
SAR	$= 71.2$ (mmoles $l^{-1})^{0.5}$

The above three examples clearly show that the SAR (71.2) when determined by using the Shahid et al. (2013) relationship (Fig. 1.2) is very close to the SAR determined by analyzing the saturation extract by modern laboratory equipment (71.2 versus 72.4). However, when SAR was determined by using the USSL Staff (1954) relationship, it was appreciably higher (95.32) than the SAR values determined by other procedures.

This indicates that the Na^+ value obtained by using the USSL Staff (1954) relationship can lead to higher SAR and can, thus, mislead the sodicity prediction. The above examples have confirmed that the relationship established for the soils of Abu Dhabi Emirate by Shahid et al. (2013) can be used reliably to determine soil sodicity (SAR and ESP). Thus, the analyses become rapid and affordable for the use in developing countries. Also of importance is the need for developing countries, e.g. the Gulf Cooperation Council (GCC) and ARASIA countries, who are relying on the USSL Staff (1954) curve to determine soluble sodium by calculating the difference between TSS and $Ca^{2+} + Mg^{2+}$, to validate this relationship for local soils.

2 Causes of Soil Salinity

There can be many causes of salts in soils; the most common sources (Plate 1.1) are listed below:

- Inherent soil salinity (weathering of rocks, parent material)
- Brackish and saline irrigation water (Box 1.1)
- Sea water intrusion into coastal lands as well as into the aquifer due to over extraction and overuse of fresh water
- Restricted drainage and a rising water-table
- Surface evaporation and plant transpiration
- Sea water sprays, condensed vapors which fall onto the soil as rainfall
- Wind borne salts yielding saline fields
- Overuse of fertilizers (chemical and farm manures)
- Use of soil amendments (lime and gypsum)
- Use of sewage sludge and/or treated sewage effluent
- Dumping of industrial brine onto the soil

Plate 1.1 Soil salinity development in agriculture and coastal fields. (**a**) Salinity in a furrow irrigated barley field, (**b**) Salinity in a sprinkler irrigated grass field, (**c**) Salinity due to sea water intrusion in coastal land

Box 1.1: Salt Loads in Soil Due to Irrigation

It is generally believed that irrigation with fresh water is safe for optimum crop production; this may be true for short duration. However, if this water is used over a long period without managing for salinity, a significant quantity of salts will be added into soil. A simple example is given below.

Assume that fresh water (EC 0.2 dS m^{-1}) is used for irrigating the crop and 8500 cubic meters per hectare (850 mm) of this water is used over the entire crop season. The water of EC 0.2 dS m^{-1} contains approximately 128 mg l^{-1} salts (0.2×640) which are equivalent to 0.128 kg per cubic meter of irrigation

(continued)

Box 1.1 (continued)
water. Over the crop season, 1088 kg of salts will, thus, be added to each hectare with the irrigation water. If we assume that the dry matter harvested from each hectare is about 15 metric tons, and there is 3.5% by weight of total salts in the harvested crop biomass, then the portion of salt harvested with the crop is 525 kg. This leaves 563 kg of salts in the soil directly or in the plant parts (belowground, stubbles, debris) which will be returned to the soil by cultivation and subsequent decay of the plant biomass left in the field. This example is a very conservative one. It is more likely that water of a higher salinity will be used for irrigation. Thus, a salinity management program needs to be implemented for virtually all irrigated agricultural crops, especially those growing in low natural rainfall areas.

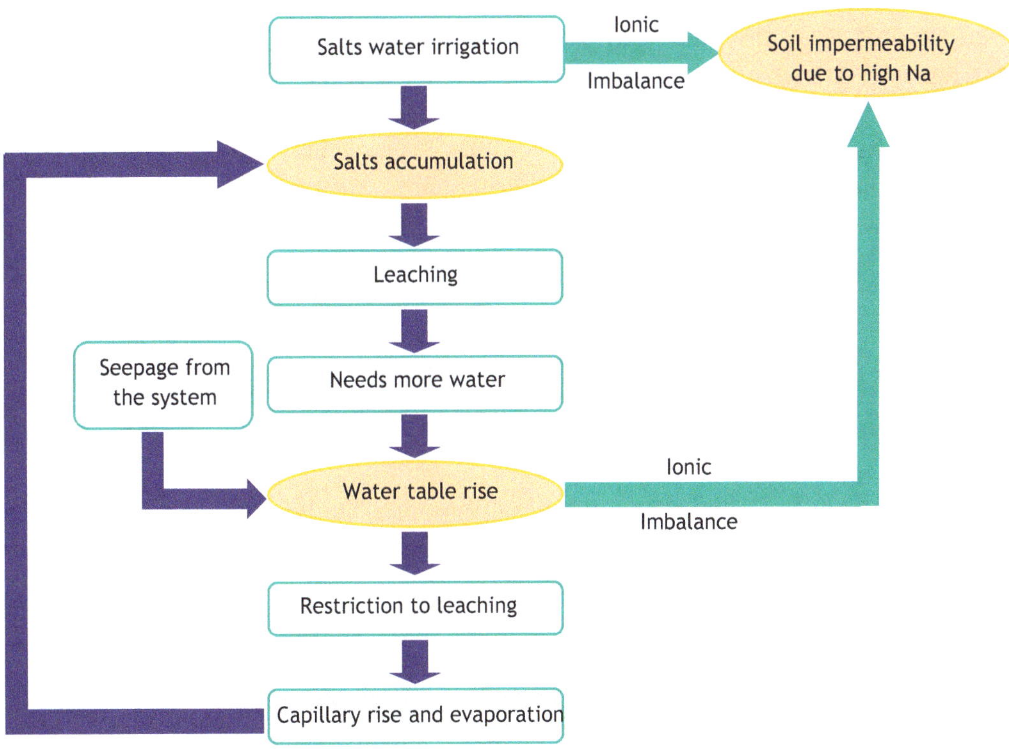

Fig. 1.5 A hypothetical soil salinization cycle. (Adapted from Shahid et al. 2010)

3 Salinity Development in Soils – A Hypothetical Cycle

Recently Shahid et al. (2010) have hypothesized a soil salinity cycle in order to present various facets in the development of soil salinity (Fig. 1.5).

4 Types of Soil Salinity

4.1 Dryland Soil Salinity

Salinity in dryland soils develops through a rising water table and the subsequent evaporation of the soil water. There are many causes of the rising water table, e.g. restricted drainage due to an impermeable layer, and when deep-rooted trees are replaced with shallow-rooted annual crops. Under such conditions, the groundwater dissolves salts embedded in rocks in the soil, with the salty water eventually reaching the soil surface, and evaporating to cause salinity. Dryland salinity can also occur in un-irrigated landscapes. There are no quick fixes to dryland salinity, though modern technologies in assessment and monitoring can allow one to follow and better understand how salinity develops. The most important of these new technologies utilizes the interpretation of remote sensing imaging over a period of time.

The potential technologies to mitigate dryland salinity include the pumping of saline groundwater and its safe disposal or use, as well as the development of alternate crop plant production systems to maximize saline groundwater use, such as deep-rooted trees. Such a deep-rooted trees system utilizes groundwater and lowers water table, the process called ***biodrainage*** (or biological drainage). In Australia dryland salinity is a major problem, one which costs Australia over $250 million a year from impacts on agriculture, water quality and the natural environment. Dryland salinity is an important problem which must be approached strategically using scientific diagnostics.

4.2 Secondary Soil Salinity

In contrast to dryland salinity, secondary salinity refers to the salinization of soil due to human activities such as irrigated agriculture.

Water scarcity in arid and desert environments necessitates the use of saline and brackish water to meet a part of the water requirement of crops. The improper use of such poor quality waters, especially with soils having a restricted drainage, results in the capillary rise and subsequent evaporation of the soil water. This causes the development of surface and subsurface salinity, thereby reducing the value of soil resource. Common ways of managing secondary salinity in irrigated agriculture are:

- Laser land leveling which facilitates uniform water distribution
- Leaching excess salts from surface soil into the subsoil
- Lowering shallow water tables with safe use or disposal of pumped saline water
- Tillage practices, seed bed preparation and seeding
- Adaption of salt tolerant plants

- Cycling the use of fresh and saline water
- Blending of fresh water with saline water
- Minimize evaporation and buildup of salts on surface soil through conservation agriculture practices such as mulching (see Chap. 6 on how nuclear techniques of oxygen-18 and hydrogen-2 help to partition between evaporation and transpiration to enhance water use efficiency on farm), addition of animal manure and crop residues, etc.

The management of secondary salinization in irrigated agriculture is discussed in more detail in Chap. 5.

5 Damage Caused by Soil Salinity

Some of the damages caused by increasing the soil salinity (Shahid 2013) are listed below:

- Loss of biodiversity and ecosystem disruption
- Declines in crop yields
- Abandonment or desertification of previously productive farm land
- Increasing numbers of dead and dying plants
- Increased risk of soil erosion due to loss of vegetation
- Contamination of drinking water
- Roads and building foundations are weakened by an accumulation of salts within the natural soil structure
- Lower soil biological activity due to rising saline water table

6 Facts About Salinity and How It Affects Plant Growth

Franklin and Follett (1985) have described the effects of salinity on plant growth as listed below:

- Proper plant selection is one way to moderate yield reductions caused by excessive soil salinity
- The stage of plant growth has a direct effect on salt tolerance; generally, the more developed the plant, the more tolerant it is to salts
- Most fruit trees are more sensitive to salts than are vegetable, field and forage crops
- Generally, vegetable crops are more sensitive to salts than are field and forage crops

7 Visual Indicators of Soil Salinity

Once soil salinity develops in irrigated agriculture fields, it is easy to see the effects on soil properties and plant growth (Plate 1.2). Visual indicators of soil salinization (Shahid and Rahman 2011) include:

- A white salt crust
- Soil surface exhibits fluffy
- Salt stains on the dry soil surface
- Reduced or no seed germination
- Patchy crop establishment
- Reduced plant vigor
- Foliage damage – leaf burn
- Marked changes in leaf color and shape occur
- The occurrence of naturally growing halophytes – indicator plants, increases
- Trees are either dead or dying
- Affected area worsens after a rainfall
- Waterlogging

8 Field Assessment of Soil Salinity

Visual assessment of salinity only provides a qualitative indication; it does not give a quantitative measure of the level of soil salinity. That is only possible through EC measurement of the soil. In the field, collection of soil saturation extract from soil paste is not possible. Therefore, an alternate procedure is used, e.g. a soil:water suspension (1:1, 1:2.5, 1:5) for field salinity assessments.

- EC can be measured on several soil:water (w/v) ratios
- EC measurement at field capacity (fc) is the most relevant representing field soil salinity. The constraint in such measurement is difficulty to extract sufficient soil water
- Compromise is EC measurement from extract collected from saturated soil paste

Plate 1.2 Field diagnostics of soil salinity – visual indicators for quick guide. (**a**) Salt stains and poor growth, (**b**) Leaf burn grassy plot, (**c**) Patchy crop establishment, (**d**) Dead trees due to salt stress

- The relationship between ECfc to ECe is generally (ECfc = 2ECe) for most of the soils, except for the sand and loamy sand textures
- Laboratory measurement of soil extract salinity (ECe) is laborious. Thus, EC of extracts using different soil:water ratios can be measured in the field and correlated to ECe, because ECe is the appropriate parameter used in salinity management and crop selection.
- Commonly used soil:water ratios in field assessment of salinity are:
  - 10 g soil +10 ml distilled water (1:1)
  - 10 g soil +25 ml distilled water (1:2.5)
  - 10 g soil +50 ml distilled water (1:5)

The EC values obtained for different soil:water ratio extracts can then be correlated to the EC of soil saturation extract (ECe) as explained below (Shahid 2013; Sonmez et al. 2008). It should be noted that EC values of 1:1, 1:2.5, 1:5 soil:water extracts are site- specific and, thus, can be used as a general guideline only. However, once correlations are established with ECe (EC of soil saturation extract) from the same soil samples, the derived ECe can be used reliably in salinity management and crop selection. Suitable conversion factors can be used based on soil type (Table 1.1).

Table 1.1 Conversion factor for deriving the ECe from the EC of extracts from different soil:water ratio suspensions

Relationship	References
ECe versus $EC_{1:1}$	
ECe = $EC_{1:1} \times 3.03$	Al-Moustafa and Al-Omran (1990) – Saudi Arabia
ECe = $EC_{1:1} \times 3.35$	Shahid (2013) – UAE (sandy soil)
ECe = $EC_{1:1} \times 3.00$	EAD (2009) – Abu Dhabi Emirate (sandy soil)
ECe = $EC_{1:1} \times 1.80$	EAD (2012) – Northern Emirates (UAE)
ECe = $EC_{1:1} \times 2.06$	Akramkhanov et al. (2008) – Uzbekistan
ECe = $EC_{1:1} \times 2.20$	Landon (1984) – Australia
ECe = $EC_{1:1} \times 1.79$	Zheng et al. (2005) – Oklahoma (USA)
ECe = $EC_{1:1} \times 1.56$	Hogg and Henry (1984) – Saskatchewan, Canada
ECe = $EC_{1:1} \times 2.7$	USSL Staff (1954) – USA
ECe = $EC_{1:1} \times 2.42$	Sonmez et al. (2008) – Turkey (sandy soil)
ECe = $EC_{1:1} \times 2.06$	Sonmez et al. (2008) – Turkey (loamy soil)
ECe = $EC_{1:1} \times 1.96$	Sonmez et al. (2008) – Turkey (clay soil)
ECe versus $EC_{1:2.5}$	
ECe = $EC_{1:2.5} \times 4.77$	Shahid (2013) – UAE (sandy soil)
ECe = $EC_{1:2.5} \times 4.41$	Sonmez et al. (2008) – Turkey (sandy soil)
ECe = $EC_{1:2.5} \times 3.96$	Sonmez et al. (2008) – Turkey (loamy soil)
ECe = $EC_{1:2.5} \times 3.75$	Sonmez et al. (2008) – Turkey (clay soil)
ECe versus $EC_{1:5}$	
ECe = $EC_{1:5} \times 7.31$	Shahid (2013) – UAE (sandy soil)

(continued)

Table 1.1 (continued)

Relationship	References
$ECe = EC_{1:5} \times 7.98$	Sonmez et al. (2008) – Turkey (sandy soil)
$ECe = EC_{1:5} \times 7.62$	Sonmez et al. (2008) – Turkey (loamy soil)
$ECe = EC_{1:5} \times 7.19$	Sonmez et al. (2008) – Turkey (clay soil)
$ECe = EC_{1:5} \times 6.92$	Alavipanah and Zehtabian (2002) – Iran (top soil)
$ECe = EC_{1:5} \times 8.79$	Alavipanah and Zehtabian (2002) – Iran (whole profile)
$ECe = EC_{1:5} \times 9.57$	Al-Moustafa and Al-Omran (1990) – Saudi Arabia
$ECe = EC_{1:5} \times 6.40$	Landon (1984) – Australia
$ECe = EC_{1:5} \times 6.30$	Triantafilis et al. (2000) – Australia
$ECe = EC_{1:5} \times 5.6$	Shirokova et al. (2000) – Uzbekistan

9 Soil Sodicity and Its Diagnostics

Sodicity is a measure of sodium ions in soil water, relative to calcium and magnesium ions. It is expressed either as sodium adsorption ratio (SAR) or as the exchangeable sodium percentage (ESP). If the SAR of the soil equals or is greater than 13 $(\text{mmoles l}^{-1})^{0.5}$, or the ESP equals or is greater than 15, the soil is termed sodic (USSL Staff 1954).

9.1 Visual Indicators of Soil Sodicity

Soil sodicity can be predicted visually in the field in the following ways

- Poorer vegetative growth than normal, with only a few plants surviving, or with many stunted plants or trees
- Variable heights of the plants
- Poor penetration of rain water – surface ponding
- Raindrop splash action – surface sealing and crusting (hard setting)
- Cloudy or turbid water in puddles
- Plants exhibit a shallow rooting depth
- Soil is often black in color due to the formation of a Na-humic substances complex
- High force required for tillage (especially in fine textured soils)
- Difficult to get soil saturation extracts in laboratory due to a filter blockage with dispersed clay

9.2 *Field Testing of Soil Sodicity*

Field assessment of relative level of soil sodicity can be determined through the use of a turbidity test on soil:water (1:5) suspensions, with ratings:

- Clear suspension – non sodic
- Partly turbid or cloudy – medium sodicity
- Very turbid cloudy – high sodicity

The relative sodicity can be further assessed by placing a white plastic spoon in these suspensions, as below.

- The spoon is clearly visible means non-sodic
- The spoon is partly visible means medium sodicity
- The spoon is not visible means high sodicity

9.3 *Laboratory Assessment of Soil Sodicity*

Accurate soil sodicity diagnostics can be made by analyzing soil samples in the laboratory. The standard presentation of soil sodicity is the exchangeable sodium percentage (ESP) derived through using sodium adsorption ratio (SAR). Alternately, ESP can be determined through measurement of exchangeable sodium (ES) and cation exchange capacity (CEC), as below.

$$\mathrm{ESP} = \left(\frac{\mathrm{ES}}{\mathrm{CEC}}\right) \times 100$$

Where, ES and CEC are represented as meq100 g^{-1}soil. An ESP of 15 is the threshold for designating soil as being sodic (USSL Staff 1954). At this ESP level, the soil structure starts degrading and negative effects on plant growth appear.

10 Sodicity and Soil Structure

A lack of sufficient volumes of fresh water for irrigation use in arid and semi-arid regions often results in the need to use water with a relatively high salinity and high sodium ion levels. It has, generally, been recognized that the sodicity affects soil permeability appreciably. The swelling and dispersion of soil clays ultimately destroys the original soil structure – likely the most important physical property affecting plant growth. The soil bulk density (the weight of soil in a given volume) and porosity (open spaces between sand, silt and clay particles in a soil) are mainly used as parameters for the soil structure. The hydraulic conductivity (the ease with which water can move through the soil pore spaces) is the net result of the effect of physical properties in the soil and is markedly affected by soil structure development. The effect of the sodicity of soil water on irrigated soils can be both a surface phenomenon, i.e. showing ***surface sealing***, as well as a subsurface phenomenon (Box 1.2), one where ***subsurface sealing*** also occurs (Shahid et al. 1992). In surface sealing, the soil water sodicity causes a breakdown and slaking of soil aggregates due to wetting. When the soil surface dries, a surface crust is formed. In subsurface sealing, the clay particles in the soil are dispersed and translocate to subsurface layers, where they are then deposited on the surface of the voids, thereby reducing void volume and blocking the pores, thus restricting further water movement, e.g. yielding non-conducting pores.

The surface sealing and crusting due to either water sodicity, or through combined effects of sodicity and raindrop splash action, have both positive and negative effects.

Box 1.2: Effect of Saline-Sodic Waters on Soil Hydraulic Conductivity and Structure in a Simulated System

In a simulated system developed and used by Shahid and Jenkins (1992a,c) for quick screening of soils with regard to their salinity and sodicity, a laboratory experiment was conducted to investigate the effect of saline-sodic water on soil structure and hydraulic conductivity (Shahid 1993). In this system, glass columns were filled with non-saline and non-sodic soil (*Typic camborthid*) which contained both swelling (smectite and vermiculite) and non-swelling (mica, chlorite and kaolinite) minerals, and silty clay loam texture. Five irrigation waters having EC 0 (deionized water), 0.5, 1.0, 1.0 and 2.4 dS m^{-1}, and SAR 0 (deionized water), 20, 25, 40 and 36 (mmoles $l^{-1})^{0.5}$, respectively, were used in wetting and drying cycles. After 14 wetting and drying cycles, the soil columns were subjected to hydraulic conductivity measurement with the respective waters treatments (above), followed by simulated rain, application of a gypsum saturated solution, and a simulated subsoil application with a gypsum saturated solution.

- The columns remained blocked with the introduction of the gypsum saturated solution. Upon examination, they revealed that a dispersion, translocation and deposition of clay platelets in conducting pores was occurring, and that this was the dominant mechanism of the much reduced hydraulic conductivity (Shahid and Jenkins 1992b; Shahid 1993).
- The columns where hydraulic conductivity was improved with gypsum saturated solution revealed, upon examination, that a swelling of clay minerals had been the main cause of hydraulic conductivity reduction.

(continued)

Box 1.2 (continued)

- The columns where hydraulic conductivity was significantly improved with gypsum saturated solution and subsoiling (disturbing soil in the column) confirmed that the dispersion, translocation and deposition of clay minerals in conducting pores was the dominant mechanism of hydraulic conductivity reduction, with swelling being a minor mechanism.
- Finally, micro-morphological observations (thin section study) of the developed soil fabric in the simulated columns revealed that the dispersion, translocation and deposition of clay platelets in the conducting pores (argillan formation) was the dominant mechanism in restricting hydraulic conductivity (Shahid 1988; Shahid and Jenkins 1991a,b).

10.1 Negative Effects of Surface Sealing

- Increased runoff particularly on slopes leading to sheet and rill erosions
- Mechanical impedance of plant seedling emergence
- Lack of aeration just below the sealed structure
- Retardation of root development
- Increased mechanical force needed for tillage (cultivation) operations

10.2 Positive Effects of Surface Sealing

- Protection against wind erosion
- More economic distribution of irrigation water since longer furrows are possible
- Protection against excessive water losses from the subsoil

11 Classification of Salt-Affected Soils

A soil which contains soluble salts in amounts in the root-zone which are sufficiently high enough to impair the growth of crop plants is defined as ***saline***. However, because salt injury depends on species, variety, plant growth stage, environmental factors, and the nature of the salts, it is very difficult to define a saline soil precisely. That said, the most widely accepted definition of a saline soil is one that has ECe more than 4 dS m^{-1} at 25 °C.

Table 1.2 Classification of salt-affected soils (USSL Staff 1954)

Soil class	ECe, dS m^{-1}	ESP	pH
Saline	≥ 4	< 15	< 8.5
Saline-sodic	≥ 4	≥ 15	≥ 8.5
Sodic	< 4	≥ 15	> 8.5

11.1 US Salinity Laboratory Staff Classification

The term ***salt-affected*** soil is being used more commonly to include saline, saline-sodic and sodic soils (USSL Staff, 1954), as summarized in Table 1.2.

11.1.1 Saline Soils

Saline soils are defined as the soils which have pHs usually less than 8.5, ECe $\geq$ 4 dS m^{-1} and exchangeable sodium percentage (ESP) < 15.

A high ECe with a low ESP tends to flocculate soil particles into aggregates. The soils are usually recognized by the presence of white salt crust during some part of the year. Permeability is either greater or equal to those of similar 'normal' soils.

11.1.2 Saline-Sodic Soils

Saline-sodic soils contain sufficient soluble salts (ECe $\geq$ 4 dS m^{-1}) to interfere with the growth of most crop plants and sufficient ESP ($\geq$ 15) to affect the soil properties and plant growth adversely, primarily by the degradation of soil structure. The pHs may be less or more than 8.5.

11.1.3 Sodic Soils

Sodic soils exhibit an ESP $\geq$15 and show an ECe < 4 dS m^{-1}. The pHs generally ranges between 8.5 and 10 and may be even as high as 11. The low ECe and high ESP tend to de-flocculate soil aggregates and, hence, lower their permeability to water.

11.1.4 Classes of Soil Salinity and Plant Growth

Electrical conductivity of the soil saturation extract (ECe) is the standard measure of salinity. USSL Staff (1954) has described general relationship of ECe and plant growth, as below.

- Non-saline (ECe ≤ 2 dS m^{-1}): salinity effects mostly negligible
- Very slightly saline (ECe 2–4 dS m^{-1}): yields of very sensitive crops may be restricted
- Slightly saline (ECe 4–8 dS m^{-1}): yields of many crops are restricted
- Moderately saline (ECe 8–16 dS m^{-1}): only salt tolerant crops exhibit satisfac tory yields
- Strongly saline (ECe >16 dS m^{-1}): only a few very salt tolerant crops show satisfactory yields

11.2 FAO/UNESCO Classification

Salt-affected soils (halomorphic soils) are also indicated on the soil map of the world (1:5,000,000) by FAO-UNESCO (1974) as solonchaks (saline) and solonetz (sodic). The origin of both terms, solonchaks and solonetz, is Russian.

11.2.1 Solonchaks (Saline)

Solonchaks (saline) are soils with high salinity (ECe >15 dSm^{-1}) within the top 125 cm of the soil.

The FAO-UNESCO (1974) divided solonchaks into four mapping units:

- ***Orthic Solonchaks***: the most common solonchaks
- ***Gleyic Solonchaks***: soils with groundwater influencing the upper 50 cm
- ***Takyric Solonchaks***: solonchaks in cracking clay soils
- ***Mollic Solonchaks***: solonchaks with a dark colored surface layer, often high in organic matter
- Soils with a lower salinity than solonchaks, but having an ECe higher than 4 dS m^{-1}, are mapped as a 'saline phase' of other soil units.

11.2.2 Solonetz (Sodic)

Solonetz (sodic) is a sodium-rich soil that has an ESP > 15. The solonetz soils are subdivided into three mapping units:

- ***Orthic Solonetz***: the most common solonetz
- ***Gleyic Solonetz***: soils with a groundwater influence in the upper 50 cm
- ***Mollic Solonetz***: soils with a dark colored surface layer, often high in organic matter

Soils with lower ESP than a solonetz, lower than 15 but higher than 6, are mapped as a 'sodic phase' of other soil units.

12 Socioeconomic Impacts of Salinity

- Reduced crop productivity on saline land leads to poverty due to income loss
- In worst case scenario, farmers abandon their land and migrate out of rural areas into urban areas which leads to unemployment
- High costs for soil reclamation, when feasible
- Loss of good quality soil (organic matter and nutrients) requires more inputs, such as fertilizer – financial pressure on farmer
- Compromised biosaline agriculture system that may give lower cash returns, relative to conventional crop production systems

13 Environmental Impacts of Salinity

- Ecosystem fragmentation
- Poor vegetative growth and cover lead to enhanced soil degradation (erosion)
- Dust with high salt levels causes environmental issues
- Sand encroaches into productive areas
- Storage capacity of water reservoirs is reduced due to eroded soil material
- Contamination of groundwater with high levels of salts occurs

14 Soil Salinity Monitoring

Soil salinity is indirectly measured as electrical conductivity of the soil solution or of a soil saturation extract. Salinity is an important analytical measurement since it reflects the suitability of the soil for growing crops. On the basis of using a soil saturation extract, ECe values of ≤ 2 dS m^{-1} (or mmhos cm^{-1}) are safe for all crops. Yields of very salt sensitive crops are negatively affected by ECe between 2 and 4 dS m^{-1}. Yields of most crops are affected by ECe between 4 and 8 dS m^{-1}. Only salt tolerant crops grow well above ECe 8 dS m^{-1}.

Salinity is largely a concern in irrigated areas and in areas with saline soils, but generally is not important in rain-fed agriculture. As use of brackish irrigation water increases, there will be greater emphasis on the utilization of soil EC measurements in the future.

Many factors can contribute to the development of saline soil conditions. However, most soils become saline through the use of salt containing groundwater for irrigation. Salt concentrations in soil vary widely, both vertically and in the

Plate 1.3 Maximum soil salinity zone in drip irrigation system in a grass field – Alain, UAE

horizontal plane. The extent of the variability depends on conditions such as differences in soil texture, growing plants which transpire soil water and also absorb salts, quality of irrigation water, soil hydraulic conductivity and the type of irrigation system being used.

A salinity monitoring plan must be an integral part of any agricultural project which deals with irrigation water which has a salinity and/or sodicity constituent. An effective salinity monitoring plan must, thus, be developed so that salinity changes can be traced, especially for the root-zone soil.

15 Soil Sampling Frequency and Zone

A number of soil sampling techniques exist, and they should be used carefully based on aims of the study. Random soil samples can be taken from a number of representative sites to get composite sample. The duration over which soil sampling occurs for salinity monitoring is quite important and that duration should be decided based on the nature of the project and its aims.

The zone of soil sampling is an important criterion, particularly for drip irrigation where the maximum salinity builds up in the periphery of the wetting front (Plate 1.3). Salt accumulation occurs via two processes.

- In the first process, the soil becomes saturated and water and solutes spread in many directions, saturating the neighboring voids before moving further.
- In the second process which occurs between consecutive irrigation cycles, both direct evaporation of water and the uptake of water and absorption of nutrients and salts by plants takes place.

Solutes are, thus, redistributed in the soil, with the final buildup of salts in the soil resulting from the interaction of the above mentioned two processes throughout the crop growth period. Soil sampling from the middle zone (between two drip irrigation lines) will, however, present maximum salinity values, and may be misleading (Plate

1.3). Therefore, soil sample from within the root-zone can provide a better estimate of the soil's salinity status.

16 Current Approaches of Salinity Diagnostics – Assessment, Mapping and Monitoring

16.1 Salinity Assessment

Accurate and reliable measurements are essential to better understand soil salinity problems in order to provide better management, improve crop yield and maintain root-zone soil health. If soil salinity can be measured, it can likely be managed. The choice of the method for soil salinity assessment, however, depends on the objective, the size of the area, the depth of soil to be assessed, number and frequency of measurements, the degree of accuracy required and the resources available.

There are a number of soil salinity assessment tools. These include salinity monitoring maps, prepared over a period of time to assess present salinity problems, and to predict future salinity risks to the area. They also include the use of salinity indicators on the soil surface, vegetation indicators, conventional salinity tests (EC 1:1 or 1:5; ECe) and more modern methods (Geophysical – EM38; Salinity sensors).

16.1.1 Routine Methods

The soil salinity measurements made using geo-referenced (using GPS) field sampling and laboratory analysis of extracts from saturated soil paste by an EC meter are accepted as the standard way of assessing soil salinity. The amount of water that a soil holds at saturation is related to soil texture, surface area, clay content, and cation exchange capacity. The lower soil:water ratios (1:1, 1:2.5, 1:5) are also used in many laboratories, though the results require calibration with ECe if they are going to be used to select salt tolerant crops.

16.1.1.1 Saturated Soil Paste – Justification for Its Use

The EC of the solution extracted from a saturated soil paste (which contains water content about twice the amount of water retained in the soil at field capacity) has been correlated with the growth or toxicity responses of a wide range of crop plants. This measure, known as electrical conductivity of the soil saturation extract (ECe), is now the generally accepted measure of soil salinity. The procedure is, however, time consuming and it requires vacuum filtration. It is important to note that EC measurements based on extracts obtained from saturated soil paste or suspensions of fixed soil:water ratios (commonly 1:1, 1:2.5 or 1:5) do not give reliable correlations. Such extracts, or extracts with a wider soil:water ratio are, however, more convenient where the ability to sample soils properly is limited. The lack of reliability is mainly due to the fact that the amount of water held at a given tension varies from soil to soil, depending on texture, the type of clay minerals present, and other factors.

16.1.1.2 Preparation of Saturated Soil Paste

- Weigh 300 g sieved (< 2 mm) air-dried soil in a 500 ml plastic beaker
- Add deionized (DI) water gradually until all the soil is moist, and mix with a spatula until a smooth paste is obtained, adding more water or more soil as necessary
- The paste should glisten and just begin to flow when the container is tilted. The saturated soil paste should have no free water on its surface, but rather should slide cleanly off a spatula
- Keep the saturated soil paste overnight with lid on the beaker
- Check the saturated soil paste the following morning by first remixing the paste and then adding water or soil as is needed to bring the paste to the saturation point – as described above

16.1.1.3 Collection of Soil Saturation Extract and EC Measurement

- Put a circular Whatman No. 42 filter paper in a Buchner funnel which is attached to a filtration rack with vacuum suction attached. Then, moisten the filter paper with DI water
- Make sure that the filter paper is tightly attached to the bottom of the funnel and that all the holes in the Buchner funnel are covered by the wet filter paper
- Start the vacuum pump, open the suction, and add the saturated soil paste to the Buchner funnel (Plate 1.4)
- Continue filtration until the soil paste on the Buchner funnel begins to develop cracks
- If the filtrate is not clear (cloudy/turbid), it should be re-filtered through another wet filter paper to obtain a clear extract. Finally, transfer the clear filtrate into a 50 ml bottle
- Switch on the conductivity meter, immerse the electrode in the soil saturation extract and record the EC reading
- Remove the conductivity cell from the filtrate, rinse it thoroughly with DI water from a squirt bottle, and carefully dry the electrode with a tissue paper

Plate 1.4 Setup for collection of extract from saturated soil paste

- If accurate comparisons of ECe are to be made across a range of samples, the temperature of the extract must be measured and a correction factor (to 25 °C) utilized. The instruments available these days automatically correct the reading to 25 °C
- Check accuracy of the EC meter using a 0.01 N KCl solution, which should give a reading of 1.413 dS m^{-1} at 25 °C

16.2 Modern Methods of Soil Salinity Measurement

16.2.1 Salinity Probe

An activity meter with a salinity probe is very convenient and gives instant apparent electrical conductivity (ECa) information, which is expressed in mS cm^{-1} and g l^{-1}. There are many models of equipment available to measure *in-situ* salinity. One is the German-made PNT3000 COMBI + model; it is commonly used in agriculture, horticulture and on landscape sites for rapid salinity assessment and monitoring. It provides an extended EC measuring range from 0 to 20 mS cm^{-1} and from 20 to 200 mS cm^{-1}. The unit includes a 250 mm long stainless steel electrode for direct soil salinity measurements; an EC-plastic probe with platinum plated ring sensors and a high quality aluminum carrying case. Its operation is convenient and simple; only one button makes the full operation possible. It is essential, however, to validate ECa values with ECe from same soil locations. In any case, the ECa must be correlated to ECe for use in assessing a crop's salt tolerance. Recently Shahid (2013) has established correlations between ECe and ECa measured by EC probe in a large number of saline fields.

ECa (Salinity Probe) *Versus* EC of Extracts (Varying Soil:Water Ratios)
For many reasons, laboratory analysis of the soil saturation extract is still the most common technique for assessing soil salinity. Salinity of the saturation extract (ECe) is considered to be the 'standard procedure' because the amount of water that a soil holds at saturation (the saturation percentage) is related to several important soil parameters, including texture, surface area, clay content, and cation exchange capacity (CEC).

Low soil to water ratios, for example 1:1; 1:2.5; 1:5, make extraction easier, but show a poorer relation to field moisture condition than the saturated paste. The choice of equipment or procedure depends upon several factors, including size of the area being evaluated, the depth of soil to be assessed, the number and frequency of the measurements, the accuracy required, and the availability of resources. The standard method of salinity monitoring involves collecting soil samples from the root-zone over a given period of time followed by their analysis in the laboratory as a soil saturation extract.

As a part of International Center for Biosaline Agriculture (ICBA) salinity monitoring program, the soil team collected a large number of soil samples from experimental plots which had been irrigated with water of varying salinity, up to a maximum of seawater (Plate 1.5). These samples were air-dried and processed in order to collect water extracts from soil:water suspensions of 1:1, 1:2.5, and 1:5, as well as soil extracts made from a saturated soil paste. The field conductivity (ECa) in mS cm^{-1} using the salinity probe (field scout) was also measured. A simple statistical test was used to calculate the correlation and correlation coefficient (R^2), and to derive factors from it in order to convert ECa (determined at several soil water contents by the use of the salinity probe) to ECe.

The correlations are developed for a fine sand (Soil Survey Division Staff 2017) textural class at ICBA experimental station (Table 1.3). The conversion factor for ECe determinations derived from the apparent EC (ECa) measured by the field scout in saline fields (Shahid 2013) was found to be: ECe = ECa × 3.81.

16.2.2 Electromagnetic Induction (EMI)

Salinity assessment and management at the farm level must help farmers improve crop productivity. The conventional field sampling followed by laboratory analysis is a tedious, expensive and time consuming process. There are other modern

Plate 1.5 Salinity monitoring in a grass field using salinity probe. (Field scout PNT 3000 COMBI +)

Table 1.3 Correlations of ECa measured by using the field scout (EC probe) with EC of 1:1, 1:2.5 and 1:5 (soil:water suspensions) and ECe

ECe =	2.2936 ECa (field scout) + 4.0177 ($R^2 = 0.8896$)
$EC_{1:1}$ =	0.7929 ECa (field scout) + 0.8131 ($R^2 = 0.9449$)
$EC_{1:2.5}$ =	0.6057 ECa (field scout) + 0.4763 ($R^2 = 0.9105$)
$EC_{1:5}$ =	0.4733 ECa (field scout) + 0.3269; ($R^2 = 0.9023$)

methods which can be used quickly and effectively in field salinity mapping, e.g. electromagnetic induction (EMI) using the EM38. EM38 is the most commonly used instrument in agricultural surveys, and gives a rapid assessment of the soil's apparent electrical conductivity (ECa), expressed in mS m^{-1}.

The EM38 has a transmitter coil and a receiving coil. The transmitter coil induces an electrical current into the soil and the receiving coil records the resultant electromagnetic field. The EM38 allows for a maximum of 150 cm or 75 cm depth of exploration in the vertical and horizontal dipole modes, respectively (Plate 1.6). EC mapping using EMI is one of the simplest and least expensive salinity measurement tools. Integration of GIS information with salinity data yields salinity maps which can help farmers interpret crop yield variations, and provide a better understanding of the subtle salinity differences across agricultural fields. These salinity maps may allow farmers to develop more precise management zones and ultimately obtain higher yields.

Plate 1.6 EC measurement with EM 38 in vertical mode (left: grassland) and horizontal mode (right: *Atriplex* field)

Technological advances such as EMI have, in the last 40 years, revolutionized soil salinity assessment. McNeill (1980) was among the first investigators to describe how the EMI method can be used in assessing soil salinity. He presented the theoretical basis for the use of EM ground conductivity meters to map lateral variation in subsurface electrical conductivity. The approach, and the instrumentation developed (GEONICS EM31, EM34-3, EM38), gained almost immediate acceptance as a replacement for the traditional use of the 4-electrode resistivity (galvanic) traversing techniques, especially when it was demonstrated that the two types of soil surveys produced very similar results (Cameron et al. 1981).

Accordingly, the spatial distribution of soil salinity on the individual field (Cameron et al. 1981), the agricultural farm (Norman et al. 1995a, b), as well as district (Vaughan et al. 1995) and the regional (Williams and Baker 1982) scales has been described. Baerends et al. (1990) used the EM38 for a detailed salinity survey in an experimental area of 37 ha. They measured electrical conductivity of the soil at more than 3600 locations, reporting significant difference between the salinity of barren, fallow and cropped fields. Thus, barren fields have a high salinity, fallow fields a moderate salinity, and the cropped fields showed the lowest salinity. Baerends et al. (1990) found good agreement between the EM38 survey and the results of the visual agronomic salinity survey, with EM38 survey yielding results with a better resolution. The EMI method is also more sensitive to salinity changes, and can be carried out at any time of the year.

The EM38 instrument has been mobilized (Rhoades 1992) by mounting it on the front wheels of a farm vehicle. The EM readings can be made with the sensor positioned at several different heights above the ground in both horizontal (EMH) and vertical (EMV) magnetic coil configurations. By this adjustment, each stop requires about 20–30 s, with multiple readings being needed to calculate ECa (and, in turn, salinity) within soil depths of 0–30, 30–60, 60–90, and 90–120 cm. The intensive data set of salinity by depth and location can also be used to assess the adequacy of past leaching/drainage practices (Rhoades 1992). For example, where salinity decreases with depth in the profile, the net flux of water (and salt) can be interpreted as being upward. This is reflective of inadequate leaching and/or poor drainage. Where salinity increases with depth in the soil profile, the net flux of water and salt can be inferred as being downward. This finding would be indicative of the adequacy of leaching/drainage. However, when salinity is low and relatively uniform with increasing depth, leaching is interpreted as excessive, probably contributing to waterlogging elsewhere, with a high salt loading of the receiving water supplies.

Rhoades and Ingvalson (1971) showed that soil salinity in the field could also be assessed with a conventional geo-electrical method. This method has been developed for the purposes of measuring the soil salinity of the entire root-zone (Nadler and Frenkel 1980; Rhoades and Oster 1986; Yadav et al. 1979). Elaborating the same technique, Rhoades and Van Schilfgaarde (1976) developed another electrical conductivity probe for measuring soil salinity distribution with increasing depth. With this probe, soil salinity is measured at specific depths in the root-zone. Thus, four electrode techniques can be successfully used for surveys of saline soils (Nadler 1981; Halvorson et al. 1977).

Williams and Baker (1982) first recognized the possibility of using EM meters for reconnaissance surveys of soil salinity variation. The high values of apparent electrical conductivity (ECa) measured by the EM meters were positively correlated with increased amounts of salts in the soil. The correlation led to empirical relationships (Rhoades et al. 1989; Cook et al. 1992; Acworth and Beasley 1998) that allow a prediction of soil salinity based on the measurement of the ECa. The correlation between salt content and the ECa is very high for zones of uniform soil material; though with zones showing different electrical conductivity, it is necessary to obtain separate values of ECa for each zone.

There is an electrical conductivity image method which provides an alternative approach. Here, the ECa function is sampled extensively in the vertical plane using a 4-electrode array. The method was described by Acworth and Griffiths (1985) and Griffiths and Baker (1993) but has not been widely used. There was difficulty in creating a starting model distribution of ECa values and extensive time consuming work was required in 'forward modeling' of the ECa data to achieve a match with the field data. Even so, the devices are used regularly for soil salinity surveys in different parts of the world (Boivin et al. 1988; Job et al. 1987; Williams and Hoey 1987). The main advantages of the EC image method are: i) measurements can be taken almost as fast as one can walk from one measurement location to another, and ii) the large volume of soil which is measured reduces the variability so that relatively fewer measurements yield a reliable estimate of the field's mean salinity.

In one study (Nettleton et al. 1994), the EM induction approach has been extended to identification of sodium-affected soils. This approach shows great promise as a measure of both salinity and sodicity. However, such studies on sodic soils have not gained much attraction due to constraints in measuring accurate sodicity levels.

Factors Affecting EC Measurement of Soil by EM38

The conductance of electricity in soil takes place through the moisture filled pores that occur between individual soil particles. Therefore, the EC of soil is determined by the following soil properties (Doerge 1999).

Porosity The greater soil porosity, the more easily electricity is conducted. Soil with high clay content has higher porosity than sandy soil. Compaction of moist soils normally increases soil EC.

Soil Water Content Dry soil is much lower in electrical conductivity than moist soil.

Salinity Level Increasing concentration of electrolytes (salts) in the soil water will dramatically increase soil EC.

Cation Exchange Capacity Mineral soils which contain high levels of organic matter (humus) and/or 2:1 clay minerals, such as montmorillonite or vermiculite, have a much higher ability to retain positively charged ions (such as Ca^{2+}, Mg^{2+}, K^+, Na^+, NH_4^+, or H^+) than the soils lacking these constituents. The presence of these cations in moisture filled soil pores will enhance soil EC in the same way as salinity does.

Temperature As temperature decreases toward the freezing point of water, soil EC decreases slightly. Below freezing point, soil pores become increasingly isolated from each other and overall soil EC declines rapidly.

16.2.3 Salinity Sensors and Data Logger

The most modern salinity data logging system is the *Real Time Dynamic Automated Salinity Logging System* (RTASLS). Here, ceramic salinity sensors are buried in the root-zone, each sensor being fitted with an external smart interface with a resolution of 16 Bits. This interface consists of an integrated microprocessor which contains all the required information to allow for autonomous operation of the sensor, including power requirements and logging interval. The smart interface is connected to a DataBus which leads to the Smart Data Logger which automatically identifies each of the salinity sensors and begins logging them at predetermined intervals. Instantaneous readings from sensors can be viewed in the field on the data logger's display. Data can also be accessed in the field with a memory stick or remotely using a smart mobile phone. Due to technology advancements, there are diversified sensors available. Such a real time data logging system (Plate 1.7) has been installed in the grass plots at the experimental station of ICBA, Dubai (Shahid et al. 2008).

A useful feature of the salinity data logging system is that it does not require knowledge of electronics or computer programing. For custom configuring the Smart Data Logger or the salinity sensors, a simple menu system can be accessed through HyperTerminal. This provides complete control over each individual sensor's setup. Data from the Smart Data Logger can be graphed using Excel.

Plate 1.7 Real time salinity logging system at ICBA experimental station (Shahid et al. 2008) (**a**) Sensor placement in the root-zone, (**b**) Buried sensors connected to the Smart Interface, (**c**) Smart Interface connected to the DataBus, (**d**) Instantaneous salinity read on Data Logger

16.2.3.1 System Installation and Operation – An Example

Salinity sensors are buried at 30 and 60 cm depths in a grass field (*Distichlis spicata* and *Sporobolus virginicus*) which is being irrigated with saline water of EC 10, 20 and 30 dS m^{-1} (Plate 1.7). The dynamic changes of soil salinity within an irrigation cycle are showing the effect of salinity of the irrigation water on the salt concentration in the grass root-zone, and how the salt levels are constantly changing under irrigation (Fig. 1.6). The soil temperature can also give assistance with interpretation of soil-water movement as no soil moisture sensors were installed. Highlights of salinity monitoring for 25 days are presented in Fig. 1.6; note that days 15–19 are a period of rain.

16.2.3.2 Soil Salinity Monitoring

The soil salinity data recorded in the *Distichlis spicata* grass field show that:

- After initial installation, it takes about 10 days for the sensors to come to equilibrium with the soil-water solution. This is especially apparent for the 30 dS m^{-1} irrigation water treatment.

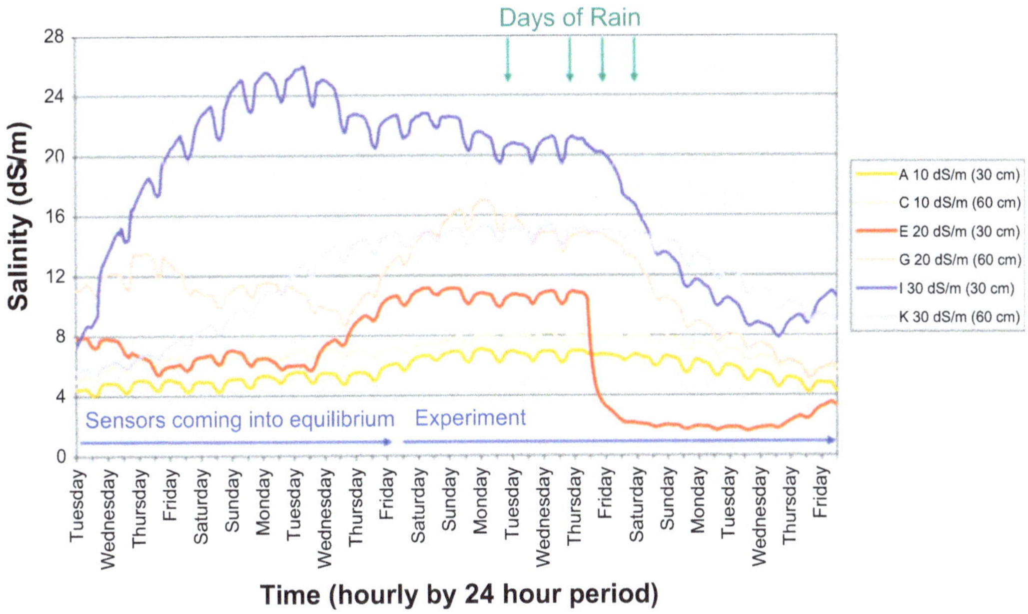

Fig. 1.6 Soil salinity monitoring in a *Distichlis spicata* field at ICBA experimental station. Y-axis shows soil salinity fluctuations on different days. The experimental plots were irrigated with saline water of EC, dS m^{-1}: 10 (A, C), 20 (E, G) and 30 (I, K)

- Salinity levels for the 10 dS m^{-1} irrigation water treatment are stable and typically 6–8 dS m^{-1}, with little change after rainfall.
- Salinity levels for the 20 dS m^{-1} irrigation water treatment are 10 dS m^{-1} at 30 cm and 14–16 dS m^{-1} at 60 cm under the standard irrigation and management practice. Rainfall rapidly reduces the salinity level at both 30 cm and 60 cm soil depths. At 60 cm, the salinity level falls by 8–10 dS m^{-1}, e.g. from 16 to 6 dS m^{-1}.
- Salinity levels for the 30 dS m^{-1} irrigation water treatment are above 20 dS m^{-1} at 30 cm and 14–16 dS m^{-1} at 60 cm under the standard irrigation and management practice. These salinity values are higher than those seen in the other treatments, reflecting the high salinity of the applied irrigation water.
- The sensitivity of the sensors to changing soil salinity levels is illustrated by both the diurnal fluctuation of salinity levels and the rapid changes that were observed (measured) after the rainfall. The diurnal data indicates a slight decline in soil salinity as the soil dries between 9:00 am and 4:00 pm, i.e. the time at which irrigation water is again applied.

16.3 Use of Remote Sensing (RS) and Geographical Information System (GIS) in Salinity Mapping and Monitoring

The RS and GIS have been used in many soil studies. Shahid (2013) has recently made efforts to compile this work in order to provide useful information to the potential users of these tools in soil salinity research. A comprehensive review is presented in the following section.

Salinity mapping can be accomplished by integrating RS and GIS techniques at both broad and small scales. The primary objective of reviewing publications dealing with these two techniques is to allow the prediction of sites vulnerable to the 'salinity menace'. GIS is a computer application that involves the storage, analysis, retrieval, and display of data that are described in terms of their geographic location. The most familiar type of spatial data is a map – GIS is really a way of storing map information electronically. A GIS map has a number of advantages over old-style maps, a primary advantage being the fact that because the data are stored electronically, they can be analyzed readily by computer. In the case of soil salinity, scientists can use data on rainfall, topography and soil type (indeed, any spatial information that is available electronically can be used) to determine the factors which make soils highly susceptible to salinization with the aim of being able to predict other (similar) regions that may be at risk.

RS imagery is well suited to map the surface expression of salinity (Spies and Woodgate 2005). For example, a poor cover of vegetation could be an indication of salinity, especially when combined with information on depth of soil to groundwater. The goal of such an exercise is to assess and map soil salinity in order to better understand the problem, and then provide information on how to take necessary actions to prevent increases in salinization of new areas over time. And finally, we need to know how to best manage salinity for sustainable use of land resources.

Salinized and cropped areas can be identified and assigned a salinity index based on greenness and brightness – one that indicates leaf moisture of plants being influenced by salinity. Here, classical false-color composites of separated bands can be used, or a computer-assisted land-surface classification can be developed (Vincent et al. 1996). A brightness index detects brightness as representing a high level of salinity. Satellite images can help in assessing the extent of saline areas and can monitor changes in real time. Saline fields are often identified by the presence of spotty white patches of precipitated salts. Such precipitates usually occur in elevated or un-vegetated areas, where evaporation has left salt residues. Such salt crusts, which can be detected on satellite images, are, however, not reliable evidence of high salinity in the root-zone. Another limitation in salinity mapping with multispectral imagery is where saline soils support productive plant growth such as areas of biosaline agriculture. There, plant cover obscures direct sensing of the soil because salt tolerant plants cannot be differentiated from other ground cover, unless extensive on-site ground investigations are made to corroborate the information.

However, RS can provide useful information for large areas with differing water and salt balances and can identify parameters such as evapotranspiration, rainfall distribution, interception losses, and crop types and crop intensities that can be used as indirect measures of salinity and waterlogging as evidence in the absence of direct estimates (Ahmad 2002).

16.4 Global Use of Remote Sensing in Salinity Mapping and Monitoring

Salinity mapping and monitoring by using RS and GIS are common in many countries and such procedures have recently been used in Kuwait and Abu Dhabi Emirate as part of the National Soil Inventories (KISR 1999a,b; Shahid et al. 2002; EAD 2009). At regional and national levels (Sukchani and Yamamoto 2005), RS and GIS was used for waterlogging and salinity monitoring (Asif and Ahmed 1999). RS technology was used for soil salinity mapping in the Middle East (Hussein 2001) and salt-affected soils using RS and GIS (Maher 1990).

The Thematic Mapper™ bands 5 and 7 are frequently used to detect soil salinity or drainage anomalies (Mulders and Epema 1986; Menenti et al. 1986; Zuluaga 1990; Vincent et al. 1996) and a broad scale monitoring of salinity using satellite remote sensing (Dutkiewics and Lewis 2008). Metternicht and Zinck (1997) have shown that Landsat TM (Thematic Mapper) and JERS-ISAR (Japan Earth Resources Satellite-Intelligent Synthetic Aperture Radar) data (visible and infrared regions) are the best ways to distinguish saline, alkaline, and non-saline soils. Landsat SMM (Solar Maximum Mission) and TM data for detailed mapping and monitoring of saline soils has been used in India for use as a reconnaissance soil map. Abdelfattah et al. (2009) developed a model that integrates remote sensing data with GIS techniques to assess, characterize, and map the state and behavior of soil salinity.

The coastal area of Abu Dhabi Emirate, where the issue of salinity is a major concern, has been used for a pilot study conducted in 2003. The development of a salinity model has been structured under four main phases: salinity detection using remote sensing data, site observations (on-site ground inspection), correlation and verification (a combination of a salinity map produced from visual interpretation of remotely sensed data and a salinity map produced from on-site observations), and model validation. Here, GIS was used to integrate the available data and information to design the model, and to create different maps. A geo-database was created and populated with data collected from observation points, melded together with laboratory analysis data. In this study, the correlation between the salinity maps developed from remote sensing data and on-site observations showed that 91.2% of the saline areas delineated using remote sensing data alone exhibit a good fit with saline areas delineated using on-site observations. Hence, the model can be adopted.

Remote sensing, thus, acquires information about the earth's surface without actually being in contact with it. The fundamentals of remote sensing in soil salinity assessment and examples of such studies from the Middle East, Kuwait, Abu Dhabi Emirate and Australia have been described recently by Shahid et al. (2010). The combination of salinity maps taken over period of time and Digital Elevation Model (DEM) help to predict salinity risk in the area.

16.5 Geo-Statistics

Geo-statistics is used for mapping of land surface features from limited sample data. It is widely used in fields where 'spatial' data is studied. Geo-statistical estimation is a two-stage process. First step is studying the gathered data to establish the predictability of values from place-to-place within the study area. This results in a graph known as a semi-variogram, which models the difference between a value at one location and the value at another location according to the distance and direction between them. The second step is estimating values at those locations which have not been sampled. This process is known as 'kriging'. The basic technique 'ordinary kriging' uses a weighted average of neighboring samples to estimate the 'unknown' value at a given location. Weights are optimized using the semi-variogram model, the locations of the samples, and all the relevant interrelationships between known and unknown values. The technique also provides a 'standard error' which may be used to calculate confidence levels. In mining, geo-statistics is extensively used in the field of mineral resource and reserve valuation, e.g. the estimation of grades and other parameters from a relatively small set of boreholes or other samples. Geo-statistics is now widely used in geological and geographical applications. However, the techniques are also used in such diverse fields as hydrology, groundwater, soil salinity mapping, and weather prediction. The application of geo-statistical techniques, such as ordinary- and co-kriging, have been applied to salinity survey data in an effort to represent more accurately the spatial distribution of soil salinity (Boivin et al. 1988; Vaughan et al. 1995).

In order to develop soil salinity maps, two approaches: Kriging and Inverse distance weighted (IDW) can be used, as explained below (Source: www.esri.com).

16.5.1 Kriging

Kriging is an advanced geo-statistical procedure that generates an estimated surface from a scattered set of points with z-values (elevation) (Fig. 1.7). Kriging assumes that the distance or direction between sample points reflects a spatial correlation that

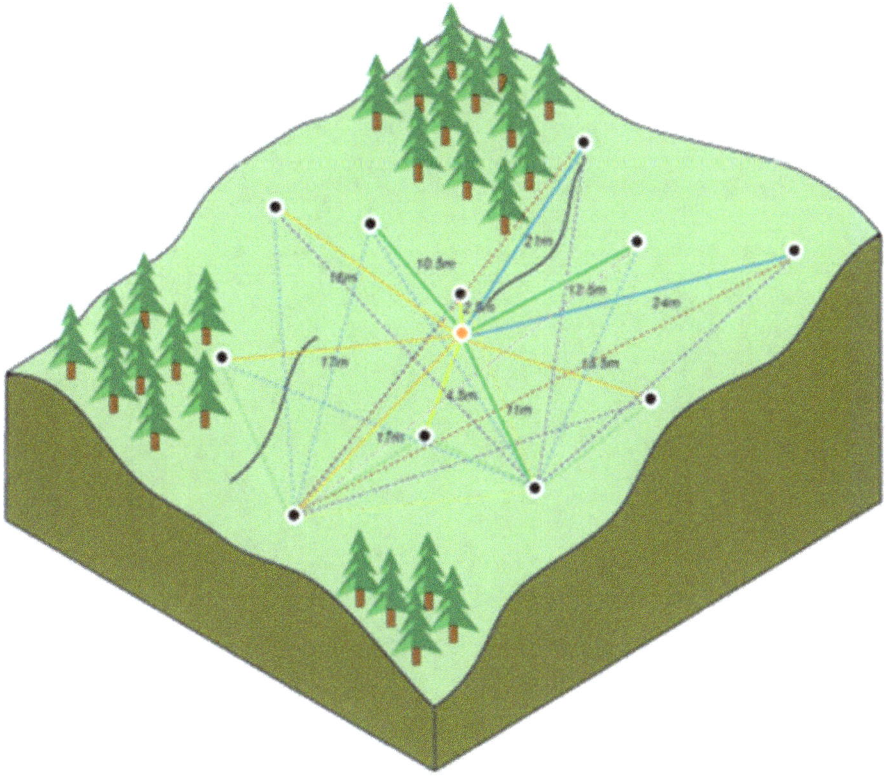

Fig. 1.7 Kriging method to generate salinity map

can be used to explain variation in the surface. The Kriging tool fits a mathematical function to a specified number of points, or to all points within a specified radius, in order to determine the output value for each location. As noted above, Kriging is a multistep process; it includes exploratory statistical analysis of the data, variogram modeling, creating the surface, and (optionally) exploring a variance surface. Kriging is most appropriate when you know there is a spatially correlated distance or directional bias in the data. It is often used in soil science and geology.

16.5.2 Inverse Distance Weighted (IDW) Interpolation

Inverse distance weighted interpolation determines cell values using a linearly weighted combination of a set of sample points (Fig. 1.8). The weight is a function of inverse distance. The surface being interpolated should be that of a location-dependent variable. This method assumes that the variable being mapped decreases in influence with distance from its sampled location.

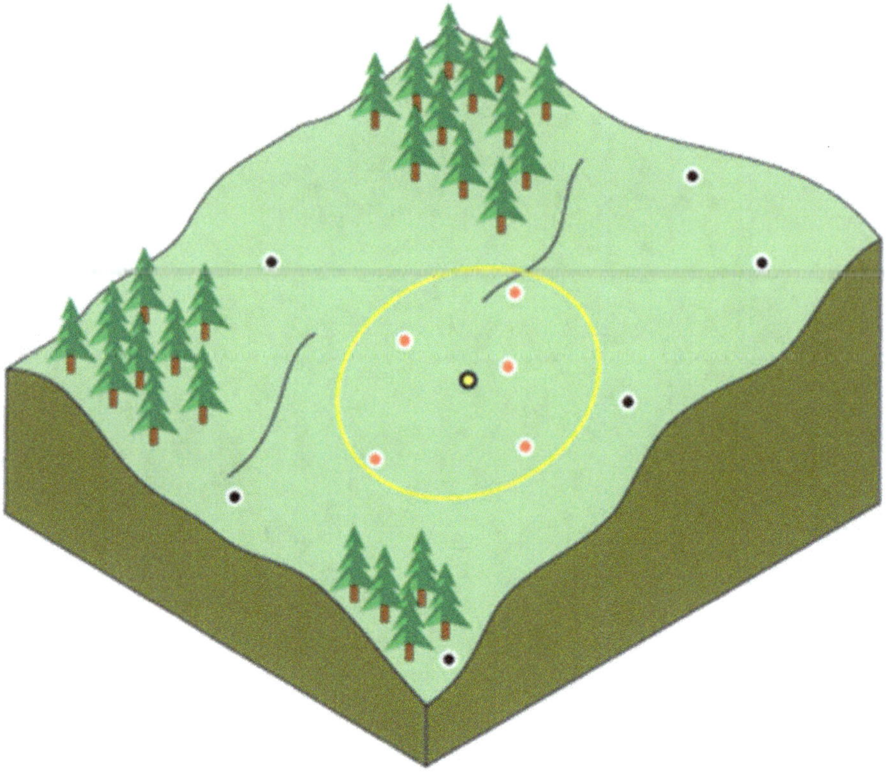

Fig. 1.8 Inverse distance weighted (IDW) method to generate salinity map

16.6 Morphological Methods

Morphological details of salt minerals can be obtained at four levels, macro-, meso-, micro- and sub-microscopic levels, as described below:

16.6.1 Macromorphology

Macromorphology can be established during field investigation on a broad scale (Plate 1.8 upper left), and features of interest can be captured with digital camera (still or video). Such investigations quickly provide useful information about the area of interest.

16.6.2 Mesomorphology

Mesomorphological observations are made when the naked eye cannot resolve the details of the features of interest (Plate 1.8 upper right). The naked eye is then aided with hand lens or binocular microscope at lower magnifications. Such observations can be made in the field (hand lens) or in laboratory using binocular low magnification microscopes.

16.6.3 Micromorphology

Micromorphology optical microscopy: When optical aid is needed for the naked eye to resolve details at higher magnifications, for example, in soil thin sections as studied with a polarizing microscope (Plate 1.8 bottom left); it is considered as an extension to field morphological studies. Soil thin sections are prepared by impregnating the saline soils with resin and hardened. After the resin has fully polymerized, the impregnated samples are cut into slices with diamond saw. The impregnated block is washed with petroleum spirit in an ultrasonic bath. One face of the impregnated soil block is then ground and polished successively with 6 and 3 μm diamond paste using Hyprez fluid as lubricant on polishing and lapping machine. The polished block is stuck to glass slide using resin and then ground to 25 μm thicknesses using lapping and polishing machine (Shahid 1988).

Micromorphology submicroscopy: Is the resolution of details at the submicroscopic level (Plate 1.8 bottom right) using an electron microscope (scanning and transmission). For submicroscopic investigation, the base of the sample and the surface of the stub are painted with a silver colloid suspension by a brush and then sample is mounted onto the stub using normal araldite. The prepared mounted samples are either carbon coated under vacuum or coated with a layer of gold-palladium (20–30 nm thick) to prevent charge buildup on the specimen, and to hold the surface of the sample and a constant electric potential. The prepared samples are then studied using the scanning electron microscopy.

The above details can be obtained as required. The microscopy techniques can also be combined. For example, Bisdom (1980) coupled optical microscopy with submicroscopic methods and later with contact microradiography (Drees and Wilding 1983). The electron microscopes can be supplemented with a micro-chemical analyzer (Energy Dispersive X Ray Analyses – EDXRA, Wavelength Dispersive X Ray Analyses – WDXRA), thereby allowing *in-situ* micro-chemical analyses (non-destructive). To study salt crusts at the submicroscopic level, soil samples need coating with gold-palladium (an alloy), or carbon applied to salt crusts is needed to make the sample a conductor. The use of submicroscopic techniques has widely been used in salt minerals studies, such as in Turkey (Vergouwen 1981) and Pakistan (Shahid 1988; Shahid et al. 1990; Shahid et al. 1992; Shahid and Jenkins 1994; Shahid 2013) (Plate 1.8).

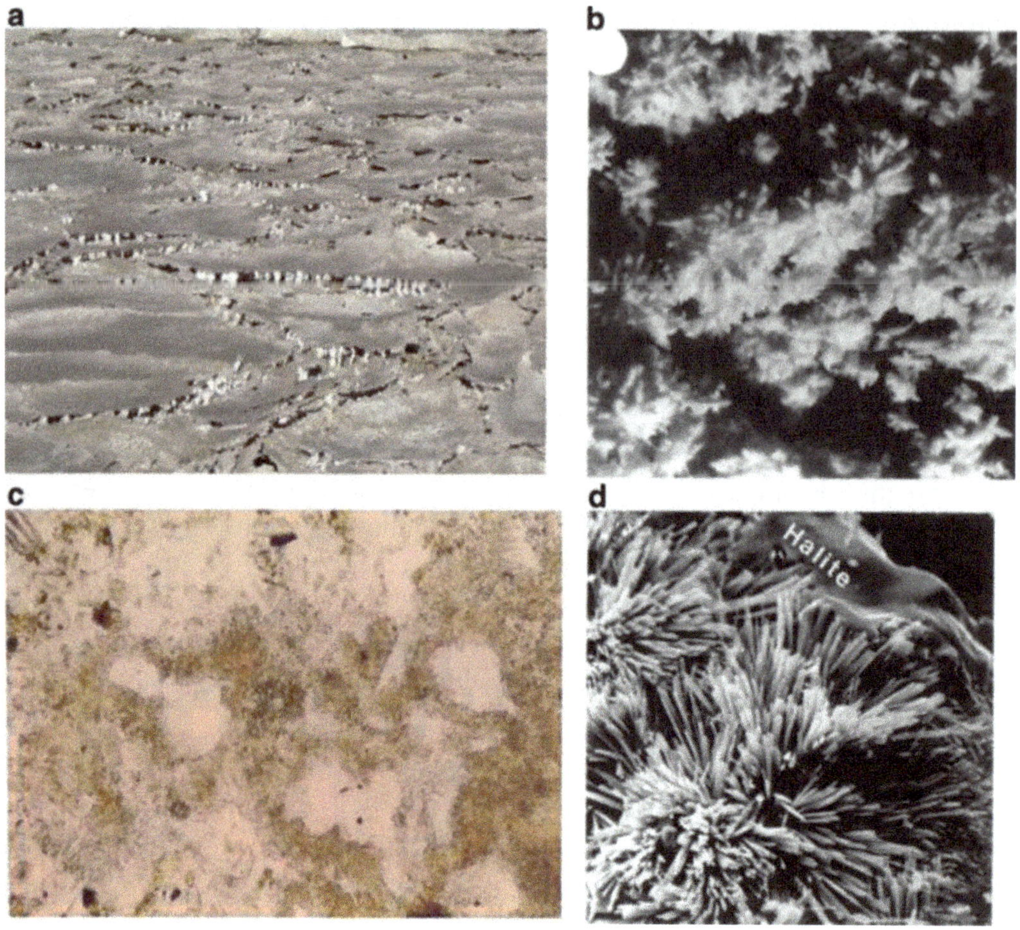

Plate 1.8 Salt features at different observations scales (cf. Shahid 1988; 2013). (**a**) Macroscopic features of soil salinity – naked eye observation (salt crust) (Width 20 meters), (**b**) Mesomorphological features – binocular observation of nahcolite (NaHCO3) mineral (Width 1100 μm), (**c**) Micromorphological feature – optical microscopic observation of thenardite (Na_2SO_4) mineral in thin section (Width 1000 μm), (**d**) Micromorphological feature – submicroscopic observation of lath shaped glauberite [($Na_2Ca\ (SO_4)_2$)] mineral – scanning electron microscopy (Width 64 μm)

References

Abdelfattah MA, Shahid SA, Othman YR (2009) Soil salinity mapping model developed using RS and GIS – a case study from Abu Dhabi, United Arab Emirates. Eur J Sci Res 26(3):342–351

Acworth RI, Beasley R (1998) Investigation of EM31 anomalies at Yarramanbah/Pump Station creek on the Liverpool Plains of New South Wales. WRL Research Report No 195. University of New South Wales, Sydney, Australia, 70 pp

Acworth RI, Griffiths DH (1985) Simple data processing of tripotential apparent resistivity measurements as an aid to the interpretation of subsurface structure. Geophys Prospect 33(6):861–867

Ahmad MD (2002) Estimation of net groundwater use in irrigated river basins using geo-information techniques: a case study in Rechna Doab, Pakistan. PhD Thesis, Wageningen University, Enschede, The Netherlands, 144 pp. http://edepot.wur.nl/121368

Akramkhanov A, Sommer R, Martius C, Hendrickx JMH, Vlek PLG (2008) Comparison and sensitivity of measurement techniques for spatial distribution of soil salinity. Irrig Drain Syst 22(1):115–126

Alavipanah SK, Zehtabian GR (2002) A database approach for soil salinity mapping and generalization from remote sensing data and geographic information system. FIG XXII International Congress, 19–26 April 2002, Washington DC, USA pp 1–8

Al-Moustafa WA, Al-Omran AM (1990) Reliability of 1:1, 1:2, and 1:5 weight extracts for expressing salinity in light-textured soils of Saudi Arabia. J King Saud Univ Agric Sci 2(2):321–329

Asif S, Ahmed MD (1999) Using state-of-the-art RS and GIS for monitoring waterlogging and salinity. International Water Management Institute, Lahore, Pakistan, 16 pp

Baerends B, Raza ZI, Sadiq M, Chaudhry MA, Hendrickx JMH (1990) Soil salinity survey with an electromagnetic induction method. Proceedings of the Indo-Pak workshop on soil salinity and water management, 10–14 February 1990, Islamabad, Pakistan, vol 1. pp 201–219

Bisdom EBA (1980) A review of the application of submicroscopy techniques in soil micromorphology. I. Transmission electron microscopy (TEM) and scanning electron microscopy (SEM). In: Bisdom EBA (ed) Submicroscopy of soils and weathered rocks. Pudoc, Wageningen, pp 67–116

Boivin P, Hachicha M, Job JO, Loyer JY (1988) Une méthode de cartographie de la salinité des sols: conductivité électromagnétique et interpolation par krigeage. Sci du sol 27:69–72

Cameron DR, De Jong E, Read DWL, Oosterveld M (1981) Mapping salinity using resistivity and electromagnetic inductive techniques. Can J Soil Sci 61:67–78

Cook PG, Walker GR, Buselli G, Potts I, Dodds AR (1992) The application of electromagnetic techniques to groundwater recharge investigations. J Hydrol 130:201–229

Doerge T (1999) Soil electrical conductivity mapping. Crop Insights 9(19):1–4

Drees LR, Wilding LP (1983) Microradiography as a submicroscopic tool. Geoderma 30:65–76

Dutkiewics A, Lewis M (2008) Broadscale monitoring of salinity using satellite remote sensing: where to from here? 2nd International Salinity Forum Adelaide, Australia, 6 pp

EAD (2009) Soil survey of Abu Dhabi Emirate. Environment Agency Abu Dhabi. 5 Volumes

EAD (2012) Soil survey of Northern Emirates. Environment Agency Abu Dhabi. 2 Volumes

FAO-UNESCO (1974) Soil map of the world. 1:5,000,000, vol 1–10. UNESCO, Paris

Franklin WT, Follett RH (1985) Crop tolerance to soil salinity. No. 505. Service in action: Colorado State University Extension Service

Griffiths DH, Baker RD (1993) Two-dimensional resistivity imaging and modeling in areas of complex geology. J Appl Geophys 29:211–226

Halvorson AD, Rhoades JD, Reule CA (1977) Soil salinity – four-electrode conductivity relationships for soils of the northern great plains. Soil Sci Soc Am J 41(5):966–971

Hogg TJ, Henry JL (1984) Comparison of 1:1 and 1:2 suspensions with the saturation extract in estimating salinity in Saskatchewan soils. Can J Soil Sci 64(4):699–704

Hussein H (2001) Development of environmental GIS database and its application to desertification study in Middle East. A Remote sensing and GIS application. PhD Thesis Graduate School of Science and Technology, Chiba University Japan, 155 pp

Job JO, Loyer JY, Ailoul M (1987) Utilisation de la conductivite electromagnetique pour la mesure directe de la salinite des sols. Cah ORSTOM, ser Pedol 23(2):123–131

KISR (1999a) Soil survey for the state of Kuwait – Volume II: Reconnaissance survey. AACM International, Adelaide

KISR (1999b) Soil survey for the state of Kuwait – Volume IV: Semi-detailed survey. AACM International, Adelaide

London JR (ed) (1984) Booker tropical soil manual, Paperback edn. Routledge Taylor & Francis Group, New York/London, 474 pp

Maher MAA (1990) The use of remote sensing techniques in combination with a geographic information system for soil studies with emphasis on quantification of salinity and alkalinity in the northern part of the Nile delta. MSc thesis Enschede Netherlands, ITC

McNeill JD (1980) Electromagnetic terrain conductivity measurements at low induction numbers. Geonics Limited Technical Notes TN-6, Geonics Ltd, Mississauga, Ontario, Canada 15 pp
Menenti M, Lorkeers A, Vissers M (1986) An application of thematic mapper data in Tunisia. ITC J 1:35–42
Metternicht GB, Zinck JA (1997) Spatial discrimination of salt- and sodium-affected soil surfaces. Int J Remote Sens 18(12):2571–2586
Mulders MA, Epema GF (1986) The thematic mapper: a new tool for soil mapping in arid areas. ITC J 1:24–29
Nadler A (1981) Field application of the four-electrode technique for determining soil solution conductivity. Soil Sci Soc Am J 45:30–34
Nadler A, Frenkel H (1980) Determination of soil solution electrical conductivity from bulk soil electrical conductivity measurements by the four-electrode method. Soil Sci Soc Am J 44:1216–1221
Nettleton WD, Bushue L, Doolittle JA, Endres TJ, Indorante SJ (1994) Sodium-affected soil identification in South-Central Illinois by electromagnetic induction. Soil Sci Soc Am J 58:1190–1193
Norman C, Challis P, Oliver H, Robinson J, Gillespie K, Rankin C, Tonkins S, Heath J (1995a) Whole farm soil salinity surveys in the Kerang region. Institute of Sustainable Irrigated Agriculture (ISIA), Tatura, Report 1993–1995, p 162
Norman C, Heath J, Turnour R, MacDonald P (1995b) On-farm salinity monitoring and investigations. Institute of Sustainable Irrigated Agriculture (ISIA), Tatura, Report 1993–1995, p 164
Rhoades JD (1992) Instrumental field methods of salinity appraisal. In: Topp GC, Reynolds WD, Green RE (eds) Advances in measurements of soil physical properties: Bringing theory into practice. SSSA Special Publication 30/ASA/CSSA/SSSA, Madison, pp 231–248
Rhoades JD, Ingvalson RD (1971) Determining salinity in field soils with soil resistance measurements. Soil Sci Soc Am Proc 35(1):54–60
Rhoades JD, Oster JD (1986) Solute content. In: Klute A (ed) Methods of soil analysis, part 1, 2nd edn. Agronomy 9:985–1006
Rhoades JD, van Schilfgaarde J (1976) An electrical conductivity probe for determining soil salinity. Soil Sci Soc Am Proc 40:647–655
Rhoades JD, Manteghi NA, Shouse PJ, Alves WJ (1989) Soil electrical conductivity and soil salinity: New formulations and calibrations. Soil Sci Soc Am J 53(2):433–439
Shahid SA (1988) Studies on the micromorphology of salt-affected soils of Pakistan. PhD Thesis, University College of North Wales, Bangor, UK. 2 Volumes
Shahid SA (1993) Effect of saline-sodic waters on the hydraulic conductivity and soil structure of the simulated experimental soil columns. Proceedings of the International Symposium on Environmental Assessment and Management of Irrigation and Drainage Projects for Sustained Agricultural Growth, 11–13 April 1993, Lahore, Pakistan, vol 2. pp 125–138
Shahid SA (2013) Developments in salinity assessment, modeling, mapping, and monitoring from regional to submicroscopic scales. In: Shahid SA, Abdelfattah MA, Taha FK (eds) Developments in soil salinity assessment and reclamation – innovative thinking and use of marginal soil and water resources in irrigated agriculture. Springer, Dordrecht/Heidelberg/New York/London, pp 3–43
Shahid SA, Jenkins DA (1991a) Effect of successive waters of different quality on hydraulic conductivity of a soil. J Drain Reclam 3(2):9–13
Shahid SA, Jenkins DA (1991b) Mechanisms of hydraulic conductivity reduction in the Khurrianwala soil series. Pak J Agric Sci 28(4):369–373
Shahid SA, Jenkins DA (1992a) Development of a simulation system for quick screening of soils against salinity and sodicity. In: Feyen J, Mwendera E, Badji M (eds) Advances in planning, design and management of irrigation systems as related to sustainable land use: proceedings of an international conference, 14–17 September 1992, vol 2. Center for Irrigation Engineering of the Katholieke Universiteit Leuven, Leuven, pp 607–614

Shahid SA, Jenkins DA (1992b) Micromorphology of surface and subsurface sealing and crusting in the soils of Pakistan. In: Vlotman WF (ed) Subsurface drainage on problematic irrigated soils: sustainability and cost effectiveness. Proceedings of the 5th international drainage workshop, 8–15 February 1992, Lahore, Pakistan, vol 2. pp 177–189

Shahid SA, Jenkins DA (1992c) Utilization of simulation system for quick screening of soils against salinity and sodicity. In: Feyen J, Mwendera E, Badji M (eds) Advances in planning design and management of irrigation systems as related to sustainable land use: proceedings of an international conference, 14–17 September 1992, vol 2. Center for Irrigation Engineering of the Katholieke Universiteit Leuven, Leuven, pp 615–626

Shahid SA, Jenkins DA (1994) Mineralogy and micromorphology of salt crusts from the Punjab, Pakistan. In: Ringrose-Voase AJ, Hymphreys GS (eds) Soil micromorphology: studies in management and genesis, developments in soil science 22, Proceedings of the IX international working meeting on soil micromorphology, July 1992, Townsville, Australia, Elsevier Amsterdam London New York Tokyo pp 799–810

Shahid SA, Rahman KR (2011) Soil salinity development, classification, assessment and management in irrigated agriculture. In: Passarakli M (ed) Handbook of plant and crop stress. CRC Press/Taylor & Francis Group, Boca Raton, pp 23–39

Shahid SA, Qureshi RH, Jenkins DA (1990) Salt-minerals in the saline-sodic soils of Pakistan. Proceedings of the Indo-Pak workshop on soil salinity and water management, 10–14 February 1990, Islamabad, Pakistan, vol 1. pp 175–191

Shahid SA, Jenkins DA, Hussain T (1992) Halite morphologies and genesis in the soil environment of Pakistan. Proceedings of the International Symposium on Strategies for Utilizing Salt-affected Lands, 17–25 February 1992, Bangkok, Thailand pp 59–73

Shahid SA, Abo-Rezq H, Omar SAS (2002) Mapping soil salinity through a reconnaissance soil survey of Kuwait and geographic information system. Annual Research Report, Kuwait Institute for Scientific Research, Kuwait, KSR 6682 pp 56–59

Shahid SA, Dakheel AH, Mufti KA, Shabbir G (2008) Automated in-situ salinity logging in irrigated agriculture. Eur J Sci Res 26(2):288–297

Shahid SA, Abdefattah MA, Omar SAS, Harahsheh H, Othman Y, Mahmoudi H (2010) Mapping and monitoring of soil salinization – remote sensing, GIS, modeling, electromagnetic induction and conventional methods – case studies. In: Ahmad M, Al-Rawahy SA (eds) Proceedings of the international conference on soil salinization and groundwater salinization in arid regions, vol 1. Sultan Qaboos University, Muscat, pp 59–97

Shahid SA, Abdelfattah MA, Mahmoudi H (2013) Innovations in soil chemical analyses: New ECs and total salts relationship for Abu Dhabi emirate soils. In: Shahid SA, Taha FK, Abdelfattah MA (eds) Developments in soil classification, land use planning and policy implications – innovative thinking of soil inventory for land use planning and Management of Land Resources. Springer, Dordrecht, pp 799–812

Shirokova YI, Forkutsa I, Sharafutdinova N (2000) Use of electrical conductivity instead of soluble salts for soil salinity monitoring in Central Asia. Irrig Drain Syst 14(3):199–205

Soil Survey Division Staff (2017) Soil survey manual. USDA-NRCS Agriculture Handbook No 18, US Government Printing Office, Washington DC, USA, 603 pp

Sonmez S, Bukuktas D, Okturen F, Citak S (2008) Assessment of different soil water ratios (1:1, 1:2.5, 1:5) in soil salinity studies. Geoderma 144(1–2):361–369

Spies B, Woodgate P (2005) Salinity mapping methods in the Australian context. Technical Report, Department of the Environment and Heritage; and Agriculture, Fisheries and Forestry, Land and Water Australia, Canberra, Australia, 234 pp

Sukchani S, Yamamoto Y (2005) Classification of salt-affected areas using remote sensing. Soil survey and classification division, Department of Land Development, Ministry of Agriculture and Cooperative, Bangkok, Thailand and Japan International Research Center Agricultural Sciences, Tsukuba, Ibaraki, Japan, 7 pp

Tanji KK (1990) Nature and extent of agricultural salinity. In: Tanji KK (ed) Agricultural salinity assessment and management, ASCE manuals and reports on engineering practice No 71. ASCE, New York, USA, pp 1–17

Triantafilis J, Laslett GM, McBratney AB (2000) Calibrating an electromagnetic induction instrument to measure salinity in soil under irrigated cotton. Soil Sci Soc Am J 64(3):1009–1017

USSL Staff (1954) Diagnosis and improvement of saline and alkali soils. USDA Handbook No 60 Washington DC, USA, 160 pp

Vaughan PJ, Lesch SM, Crown DL, Cone DG (1995) Water content effect on soil salinity prediction. A geostatistical study using co-kriging. Soil Sci Soc Am J 59(4):1146–1156

Vergouwen L (1981) Scanning electron microscopy applied on saline soils from the Konya basin in Turkey and from Kenya. In: Bisdom EBA (ed) Submicroscopy of soil and weathered rocks. Center for Agricultural Publishing and Documentation, Wageningen, pp 237–248

Vincent B, Vidal A, Tabbet D, Baqri A, Kapur M (1996) Use of satellite remote sensing for the assessment of waterlogging or salinity as an indication of the performance of drained systems. In: Vincent B (ed) Evaluation of performance of subsurface drainage systems. Proceedings of the 16th congress on irrigation and drainage, Cairo, ICID New Delhi, India pp 203–216

Williams BG, Baker GC (1982) An electromagnetic induction technique for reconnaissance survey of salinity hazards. Aust J Soil Res 20(2):107–118

Williams BG, Hoey D (1987) The use of electromagnetic induction to detect the spatial variability of the salt and clay contents of soils. Aust J Soil Res 25(1):21–27

Yadav BR, Rao NH, Paliwal KV, Sharma PBS (1979) Comparison of different methods for measuring soil salinity under field conditions. Soil Sci 127(26):335–339

Zheng H, Schroder JL, Pittman JJ, Wang JJ, Payton ME (2005) Soil salinity using saturated paste and 1:1 soil to water extracts. Soil Sci Soc Am J 69(4):1146–1151

Zuluaga JM (1990) Remote sensing applications in irrigation management in Mendoza, Argentina. In: Menenti M (ed) Remote sensing in evaluation and management of irrigation. Instituto Nacional de Ciencia y Tecnic, Mendoza, pp 37–58

Chapter 2
Soil Salinity: Historical Perspectives and a World Overview of the Problem

Shabbir A. Shahid, Mohammad Zaman, and Lee Heng

Abstract Soil salinity is not a recent phenomenon, it has been reported since centuries where humanity and salinity have lived one aside the other. A good example is from Mesopotamia where the early civilizations first flourished and then failed due to human-induced salinization. A publication '*Salt and silt in ancient Mesopotamian agriculture*' highlights the history of salinization in Mesopotamia where three episodes (earliest and most serious one affected Southern Iraq from 2400 BC until at least 1700 BC, a milder episode in Central Iraq occurred between 1200 and 900 BC, and the east of Baghdad, became salinized after 1200 AD) have been reported. There are reports clearly revealing that '*many societies based on irrigated agriculture have failed*', e.g. Mesopotamia and the Viru valley of Peru. The flooding, over-irrigation, seepage, silting, and a rising water table have been reported the main causes of soil salinization. Recent statistics of global extent of soil salinization do not exist, however, various scientists reported extent differently based on different data sources, such as there have been reports like, 10% of the total arable land as being affected by salinity and sodicity, one billion hectares are covered with saline and/or sodic soils, and between 25% and 30% of irrigated lands are salt-affected and essentially commercially unproductive, global distribution of salt-affected soils are 954 million ha, FAO in 1988 presented 932 million ha salt-affected soils, of almost 1500 million ha of dryland agriculture, 32 million ha are salt-affected. Precise information on the recent estimates of global extent of salt-affected soils do not exist, many countries have assessed their soils and soil salinization at the national level, such as Kuwait, United Arab Emirates, Middle East, and Australia etc. Considering the current extent of salt-affected soils the cost of salt-induced land degradation in 2013 was $441 per hectare, a simple benefit transfer suggests the current annual economic losses could be $27 billion.

Keywords Historical perspective · Mesopotamia · Iraq · Global extent · Economic losses · Viru valley

M. Zaman et al., *Guideline for Salinity Assessment, Mitigation and Adaptation Using Nuclear and Related Techniques*, https://doi.org/10.1007/978-3-319-96190-3_2

1 Introduction

Soil salinity is a major global issue owing to its adverse impact on agricultural productivity and sustainability. Salinity problems occur under all climatic conditions and can result from both natural and human-induced actions. Generally speaking, saline soils occur in arid and semi-arid regions where rainfall is insufficient to meet the water requirements of the crops, and leach mineral salts out of the root-zone. The association between humans and salinity has existed for centuries and historical records show that many civilizations have failed due to increases in the salinity of agricultural fields, the most known example being Mesopotamia (now Iraq). Soil salinity undermines the resource base by decreasing soil quality and can occur due to natural causes or from misuse and mismanagement to an extent which jeopardizes the integrity of soil's self-regulatory capacity.

Soil salinity is dynamic and spreading globally in over 100 countries; no continent is completely free from salinity (Fig. 2.1). Soil salinization is projected to increase in future climate change scenarios due to sea level rise and impact on coastal areas, and the rise in temperature that will inevitably lead to increase evaporation and further salinization. Salinization of soils can affect ecosystems to an extent where they no longer can provide 'environmental services' to their full potential. It is realized that recent estimates of the global extent of soil salinization do not exist. But it can be assumed that, since the earlier data gathering in the 1970s and 1980s, salinization has expanded as newly affected areas most probably exceed the areas restored through reclamation and rehabilitation. There is a long list of countries where salt-induced land degradation occurs. Some well-known regions where salinization is extensively reported include the Aral Sea Basin (Amu-Darya and Syr-Darya River Basins) in Central Asia, the Indo-Gangetic Basin in India, the

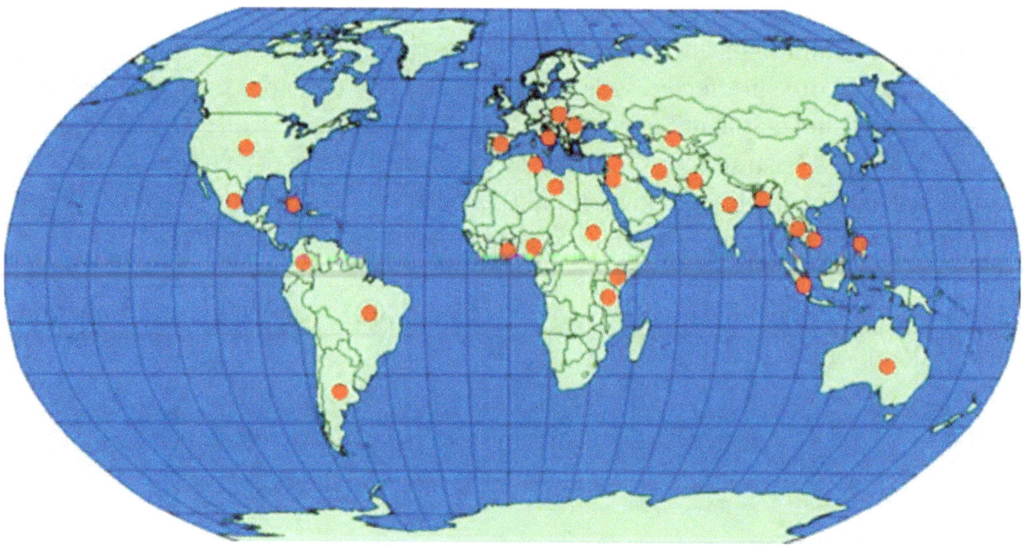

Fig. 2.1 World map representing countries with salinity problems. (https://www.researchgate.net/publication/262495450)

Indus Basin in Pakistan, the Yellow River Basin in China, the Euphrates Basin in Syria and Iraq, the Murray-Darling Basin in Australia, and the San Joaquin Valley in the United States (Qadir et al. 2014).

2 Soil Salinity – A Historical and Contemporary Perspective

For centuries, humanity and salinity have lived one aside the other. There is good evidence for Mesopotamia that early civilizations flourished and then failed due to human-induced salinization. Jacobson and Adams (1958), in their publication, '*Salt and silt in ancient Mesopotamian agriculture*' highlighted the history of salinization in Mesopotamia. Ancient records show three episodes of soil salinization in Iraq. The earliest and the most serious one affected Southern Iraq from 2400 BC until at least 1700 BC. A milder episode in Central Iraq occurred between1200 BC and 900 BC, and there is archeological evidence that the Nahrwan area, east of Baghdad, became salinized after 1200 AD. Flooding, over-irrigation, seepage, silting, and a rising water-table are considered to be the main reasons for these episodes of increased salinization (Gelburd 1985).

In southern Iraq in 3500 BC, both wheat and barley were equally important cultivated crops, though after 100 years wheat had slipped to one sixth, and by 2100 BC, its cultivation had become almost insignificant, dropping to only 2%. By 1700 BC, wheat cultivation was completely phased out. Historical records show that concurrent with the shift to barley cultivation, there was an appreciable and serious decline in soil fertility and gradual declines in barley yields, which for the most part can be attributed to salinization (Jacobson and Adams 1958). Thus, after almost 5000 years of successful irrigated agriculture, the Sumerian civilization failed. In the Indus plains of Pakistan and India, the practice of irrigation began about 2000 years ago during the Harapa civilization, but it is only recently that serious problems of salinity and sodicity are being encountered. In the Viru valley of Peru, irrigated agriculture began between 800 BC and 30 AD (Wiley 1953), and by 800 AD, the population was at a peak. Then from 1200 AD, the population declined appreciably and the residents of the once densely populated Viru valley bottom relocated to more narrow upper reaches of the valley. The historians partly attribute this relocation to increased salinity and a rising water- table, together with inadequate soil drainage (Armillas 1961). Tanji (1990) draws a historical perspective of irrigation-induced salinity in several regions. In a similar perspective, general remarks by Wiley (1953) clearly reveal that '*many societies based on irrigated agriculture have failed*', e.g. Mesopotamia and the Viru valley of Peru.

3 An Overview of Salinity Problem

The earth's land surface is 13.2×10^9 ha, but only 7×10^9 ha of this is arable, of which only 1.5×10^9 ha is currently cultivated (Massoud 1981). Of the cultivated lands, about 0.34×10^9 ha (23%) are saline and 0.56×10^9 ha (37%) are sodic. Older estimates (Szabolcs 1989) described 10% of the total arable land as being affected by salinity and sodicity, with the effects extending to over 100 countries in all continents. One billion hectares of the 13.2×10^9 hectares of the land is, thus, covered with saline and/or sodic soils, and between 25% and 30% of irrigated lands are salt-affected and essentially commercially unproductive.

The countries affected by salinization are predominantly located in arid and semi-arid regions, where continued irrigation with low quality groundwater has taken place. Low rainfall has also contributed to the expansion of salt-affected soils. The largest area of the world's saline soils occurs in the arid and semi-arid regions (Massoud 1974; Ponnamperuma 1984), where evapotranspiration exceeds precipitation. The rapid conversion into barren land through salinity/sodicity has negatively affected the environment and has substantially altered natural resources in a number of countries. Worldwide, some ten million hectares of irrigated land is abandoned annually because of salinization, sodication and waterlogging (Szabolcs 1989). These degraded soils occur principally in the hot arid and semi-arid regions, although they have also been recorded in Polar Regions (Buringh 1979).

Global statistics on salt-affected soils vary according to different data sources. Saline soils occupied more than 20% of the world's irrigated area by the mid-1990s (Ghassemi et al. 1995). Since then, the extent of salinity has likely increased and, in some countries, salt-affected soils occur on more than half of the irrigated lands (Metternicht and Zinck 2003). Kovda and Szabolcs (1979) reported global distribution of salt-affected soils as 954 million ha. Data summarized from Szabolcs (1974) for Europe and Massoud (1977) for the other continents, as reported by Abrol et al. (1988) in FAO Soils Bulletin 39, presents 932.2 million ha of salt-affected soils (Table 2.1). Of almost 1500 million ha of dryland agriculture, 32 million ha are salt-affected (FAO 2000). Although recent estimates of global extent of salt-affected soils do not exist, many countries have assessed their soils and soil salinization levels at the national level, such as Kuwait (Shahid et al. 2002), United Arab Emirates

Table 2.1 Worldwide distribution of salt-affected areas (Million ha)

Area	Saline soils	Sodic soils	Total	Percent
Australasia	17.6	340.0	357.6	38.4
Asia	194.7	121.9	316.5	33.9
America	77.6	69.3	146.9	15.8
Africa	53.5	26.9	80.4	8.60
Europe	7.8	22.9	30.8	3.30
World	**351.2**	**581.0**	**932.2**	**100**

Source: Abrol et al. (1988) in FAO Soil Bulletin 39; Summary of data for Europe (Szabolcs 1974) and for other continents (Massoud 1977)

(EAD 2009, 2012), Middle East (Hussein 2001; Shahid et al. 2010), and Australia (Oldeman et al. 1991).

In a comprehensive overview of the global identification of salinity problems and the global salinity status, Shahid (2013) reported that about 200 × 10^6 hectares of land is affected by salinity in Southwest USA and Mexico. In Spain, Portugal, Greece, and Italy, saltwater intrusion into aquifers is appreciable; in Spain more than 20% of the land area is desert, or is seriously degraded and, thus, nonproductive.

In the Middle East, 20 × 10^6 hectares are affected by increased soil salinity, the reasons being poor irrigation practices, high evaporation rates, growth of ***sabkhas*** (salt scalds), and an increase in groundwater salinity. In addition, productivity of the irrigated lands of the Euphrates basin (Syria, Iraq) is seriously constrained by salinity. In Iran, 14.2% of the total land area is salt-affected (Pazira 1999). In Egypt, 1 × 10^6 hectares of land which could be cultivated along the Nile is salt-affected. Salt accumulation in the Jordan River basin adversely affects agricultural production in Syria and Jordan. In Africa, 80 × 10^6 hectares is saline, sodic, or saline-sodic, of which the Sahel, in West Africa, is the most affected.

In Asia, 20% of India's cultivable land, mainly in Rajasthan, coastal Gujarat, and the Indo-Gangetic plains, is affected by salinity or sodicity. In Pakistan, 10 × 10^6 hectares is affected and about 5–10 hectares per hour is lost to salinity and/or waterlogging in coastal regions and in the irrigated Indus basin. In Bangladesh, 3 × 10^6 hectares is unproductive due to salinity. In Thailand, 3.58 × 10^6 hectares is salt-affected (3.0 × 10^6 hectares being inland and 0.58 × 10^6 hectares being coastal saline soils). In China, 26 × 10^6 hectares of their total land area is salt-affected (Inner Mongolia, the Yellow River basin and tidal coastal regions), while in Australia the extent of saline soils is 357 × 10^6 hectares.

The global extent and distribution of 76.6 million hectares of human-induced salt-affected soils (Oldeman et al. 1991) and a similar distribution for irrigated lands affected by secondary salinization (Ghassemi et al. 1995) are presented in Tables 2.2 and 2.3. These soils are distributed in desert and semi-desert regions, frequently occurring in fertile alluvial plains, river valleys, coastal areas and in irrigation districts. The countries where significant salinity problems exist include, but are not limited, to Australia, China, Egypt, India, Iran, Iraq, Mexico, Pakistan, the

Table 2.2 Global extent of human-induced salinization (Oldeman et al. 1991; Mashali 1995)

Continent	Degree of salinization and affected area (mha)					Percent
	Light	Moderate	Strong	Extreme	Total	
Africa	4.7	7.7	2.4	–	14.8	19.3
Asia	26.8	8.5	17.0	0.4	52.7	68.8
South America	1.8	0.3	–	–	2.1	2.7
North & central America	0.3	1.5	0.5	–	2.3	3.0
Europe	1.0	2.3	0.5	–	3.8	5.0
Australia	–	0.5	–	0.4	0.9	1.2
World total	**34.6**	**20.8**	**20.4**	**0.8**	**76.6**	**100**

Table 2.3 Global estimates of secondary salinization in the world's irrigated lands. (Summarized from Ghassemi et al. 1995; Mashali 1995)

	Area (mha)		
Country	Cropped	Irrigated	Salt-affected[a]
China	97.0	44.8	6.7 (15)
India	169.0	42.1	7.0 (17)
Commonwealth of independent states	232.5	20.5	3.7 (18)
United States of America	190.0	18.1	4.2 (23)
Pakistan	20.8	16.1	4.2 (26)
Iran	14.8	5.8	1.7 (29)
Thailand	20.0	4.0	0.4 (10)
Egypt	2.7	2.7	0.9 (33)
Australia	47.1	1.8	0.2 (11)
Argentina	35.8	1.7	0.6 (35)
South Africa	13.2	1.1	0.1 (9)
Subtotal	**842.9**	**158.7**	**29.7 (19)**
World (Total)	**1474**	**227**	**45 (20)**

[a]Salt-affected soils within the irrigated area; values in parentheses are percentage

USSR, Syria, Turkey, and the United States. In Gulf States (Bahrain, Kuwait, Saudi Arabia, Qatar, Oman, and the United Arab Emirates), saline soils mainly occur in coastal lands (due to seawater intrusion), and also on agricultural farms irrigated with saline/brackish water.

Secondary salinization (i.e., soil salinization due to human activities such as irrigated agriculture) is predominantly located in the arid and semi-arid regions including Egypt, Iran, Iraq, India, China, Chile, Argentina, Commonwealth of Independent States, Spain, Thailand, Pakistan, Syria, Turkey, Algeria, Tunisia, Sudan and the Gulf States. About 76.6 million hectares (Table 2.2) of cultivated lands are salt-affected by human-induced processes (Oldeman et al. 1991; Mashali 1995; Ghassemi et al. 1995) and approximately 30 million ha can be attributed to secondary salinization of non-irrigated lands. However, according to Ghassemi et al. (1995), globally 20% or 45 million hectares out of a total 227 million hectares of irrigated land are salt-affected (Table 2.3).

4 Distribution of Salinity in Drylands in Different Continents of the World

As reported by UNEP (1992), the distribution of salt-affected soils in drylands in different continents is presented in Table 2.4. These soils are divided into two categories: saline (412 million hectares) and sodic (618 million hectares), totaling 1030 million hectares. Australasia has the widest distribution with 357.6 million hectares, followed by Africa with 209.6 million hectares.

Table 2.4 Salt-affected soils in drylands by continents (UNEP 1992; cf FAO-ITPS-GSP 2015)

Continent	Salt-affected area (mha)		
	Saline soils	Sodic soils	Total
Africa	122.9	86.7	209.6
Australasia	17.6	340.0	357.6
Mexico/Central America	2.0	–	2.0
North America	6.2	9.6	15.8
North and Central Asia	91.5	120.2	211.7
South America	69.5	59.8	129.3
South Asia	82.3	1.8	84.1
Southeast Asia	20.0	–	20.0
Total	**412.0**	**618.1**	**1030.1**

Table 2.5 Soil salinity caused by irrigation in major irrigating countries and in the world (Postel 1989)

Country	Area damaged	
	mha	% of irrigated land
India	20.0	36.0
China	7.0	15.0
United States of America	5.2	27.0
Pakistan	3.2	20.0
Soviet Union	2.5	12.0
Total	37.9	24.0
World	**60.2**	**24.0**

5 Irrigation Practices and Soil Salinization

The practice of irrigation, if not planned and managed properly, can result in increased soil salinization. An estimate (Postel 1989) shows that about 25% of the world's irrigated lands are damaged by salinity, while Adams and Hughes (1990) have reported that up to 50% of irrigated lands are affected by salt. Szabolcs (1989) states that no continent is free from salt-affected soils and serious salt-related problems occur in at least 70 countries. Table 2.5 shows the area of irrigated land damaged by salinization for the five worst-affected countries (Postel 1989).

6 Regional Overview of Salinity Problem

More recent estimates of the regional distribution of saline soils do not exist. There is a need to update this information in order to better understand the extent of the problem and to develop soil use and management policies. Such estimates are essential given the continuing decline of soil resources for food production. An earlier search of the literature (Mashali 1995; FAO-Unesco Soil map of the world

Table 2.6 Regional distribution of salt-affected soils (mha). (cf. Mashali 1995)

Region	Solonchak – saline phase	Solonetz – sodic phase	Total
North America	6	10	16
Mexico and Central America	2	–	2
South America	69	60	129
Africa	54	27	81
South and West Asia	83	2	85
South East Asia	20	–	20
North and Central Asia	92	120	212
Australasia	17	340	357
Europe	9	21	30
Total	**352**	**580**	**932**

1974) does, however, give an estimate of the extent of the regional distribution of salt-affected soils (Table 2.6). These estimates show the total extent to be 932 million hectares of salt-affected lands, with the maximum area occurring in the region of Australasia (357 million ha).

7 Extent of Soil Salinity in the Middle East

Information regarding the extent of salinization in the Middle East is very limited. However, some general information has been obtained through the use of Remote Sensing imagery and other methods. This information was used to develop a soil salinization map of the Middle East (Hussein 2001; Shahid et al. 2010). In this map, salinization was divided into four general categories: slight, moderate, severe and very severe, as shown in Table 2.7. Earlier, an estimated area of 209,000 hectares has been reported as being salinized in Kuwait (Hamdallah 1997), which is roughly 3% of the total Kuwait land area.

Table 2.7 shows an area of 11.2% of the Middle East being affected to varying levels by soil salinization. Realizing the soil salinity, a hazard to agriculture and to the ecosystem services, Shahid et al. (1998) described soil salinization as early warning of land degradation in Kuwait. Later, Shahid et al. (2002) interpreted the soil survey data (KISR 1999) using GIS and mapped soil salinity into different salinity zones, where area occupied by each zone is as: 4.1 – 10 dS m^{-1} (0.685%), 10.1 – 25.0 dS m^{-1} (4.37%), and more than 25 dS m^{-1} (7.06%). This concludes an area of about 12.1% affected to varying degrees of salinity in the entire state of Kuwait. In the Abu Dhabi Emirate (EAD 2009), an area of 35.5% (2,034,000 ha) has been depicted to be affected to varying degrees of soil salinity. The highly saline soils on the soil salinity map are confined to the coastal land (King et al. 2013), the areas of deflation plains, and inland ***sabkha*** (salt scald) where the groundwater levels approach the surface, creating large areas of aquisalids at the great group level of US soil taxonomy (Soil Survey Staff 2014; Shahid et al. 2014).

Table 2.7 Salinization classes and affected area in the Middle East (Hussein 2001; Shahid et al. 2010)

Class	Area km^2	Area %
Slight	113,814	1.72
Moderate	109,148	1.65
Severe	380,025	5.74
Very severe	138,204	2.09
Total	**741,191**	**11.2**

8 Socioeconomic Aspects of Soil Salinization

A comprehensive review of published literature revealed very few publications dealing with socioeconomic aspects of salt-induced land degradation. On the global level, generation of such information requires appreciable resources and the commitment of properly trained staff to the project. However, Qadir et al. (2014) conclude that previous studies show a limited number of highly variable estimates of the costs of salt-induced land degradation. Even so, they have made simple extrapolations from these studies and the estimates show that the global annual cost of salt-induced land degradation in irrigated areas could be US$ 27.3 billion in lost crop production. Based on these estimates, Qadir et al. (2014) recommended investing in the remediation of salt-affected lands and noted that remediation costs must be included in a broader national strategy for food security, and defined in national action plans.

Qadir et al. (2014) identified countries where such economic cost on salt-induced soil degradation has been reported, including but not necessarily limited to Australia, India, the United States, Iraq, Pakistan, Kazakhstan, Uzbekistan, and Spain. They further indicated that the valuation of the cost of salt-induced land degradation has been mainly based on estimates of crop production losses. However, it is unclear whether their comparisons are made with crop production values taken from land not affected by salinity.

Taking the above examples into account, Qadir et al. (2014) have concluded that, considering the current extent of global irrigated area 310 million hectares (FAO-AQUASTAT 2013) and 20% of this area as salt-affected (62 million hectares), and the inflation-adjusted cost of salt-induced land degradation in 2013 as US$ 441 per hectare, a simple benefit transfer suggests the current annual economic losses could be US$ 27.3 billion.

References

Abrol IP, Yadav JSP, Massoud FI (1988) Salt-affected soils and their management. FAO Soils Bulletin 39, Food and Agriculture Organization of the United Nations, Rome, Italy, 131 pp

Adams WM, Hughes FMR (1990) Irrigation development in desert environments. In: Goudi AS (ed) Techniques for desert reclamation. Wiley, New York, pp 135–160

Armillas P (1961) Land use in pre-Columbian America. In: Stamp LD (ed) A history of land use in arid regions, vol 17. UNESCO Arid Zone Research, Paris, pp 255–276

Buringh P (ed) (1979) Introduction to the soils of tropical and subtropical regions, 3rd edn. Center for Agricultural Publishing and Documentation, Wageningen, 124 pp

EAD (2009) Soil survey for the emirate of Abu Dhabi. Reconnaissance soil survey report. Volume 1. Environment Agency Abu Dhabi, United Arab Emirates

EAD (2012) Soil survey of northern emirates. Environment Agency Abu Dhabi, 2 Volumes

FAO (2000) Extent and causes of salt-affected soils in participating countries. Global network on integrated soil management for sustainable use of salt-affected soils. FAO-AGL website

FAO-AQUASTAT (2013) Area equipped for irrigation and percentage of cultivated land. Available at http://www.fao.org/nr/water/aquastat/globalmaps/index.stm. Accessed 16 Sept 2013

FAO-ITPS-GSP (2015) Status of the world's soil resources. FAO-ITPS-GSP Main Report, Food and Agriculture Organization of the United Nations, Rome, Italy, pp 125–127

FAO-UNESCO (1974) FAO-Unesco soil map of the world. 1:5 000 000, UNESCO Paris, 10 volumes

Gelburd DE (1985) Managing salinity lessons from the past. J Soil Water Conserv 40(4):329–331

Ghassemi F, Jakeman AJ, Nix HA (1995) Salinisation of land and water resources: human causes, extent, management and case studies. CABI Publishing, Wallingford, 526 pp

Hamdallah G (1997) An overview of the salinity status of the near east region. Proceedings of the workshop on management of salt-affected soils in the Arab Gulf States, Abu Dhabi, United Arab Emirates, 29 October–2 November 1995. Food and Agriculture Organization of the United Nations, Regional Office for the Near East, Cairo, Egypt, pp 1–5

Hussein H (2001) Development of environmental GIS database and its application to desertification study in Middle East – a remote sensing and GIS application. PhD thesis Graduate School of Science and Technology, Chiba University Japan, 155 pp

Jacobson T, Adams RM (1958) Salt and silt in ancient Mesopotamian agriculture. Science (New Series) 128(3334):1251–1258

King P, Grealish G, Shahid SA, Abdelfattah MA (2013) Land evaluation interpretations and decision support systems: soil survey of Abu Dhabi emirate. Chapter 6. In: Shahid SA, Taha FK, Abdelfattah MA (eds) Developments in soil classification, land use planning and policy implications-innovative thinking of soil inventory for land use planning and management of land resources. Springer, Dordrecht, pp 147–164

KISR (1999) Soil survey for the state of Kuwait – volume IV: semi-detailed survey. AACM International, Adelaide

Kovda VA, Szabolcs I (eds) (1979) Modelling of soil salinization and alkalization: supplementum. Vol 28 of Agrokemia es Talajtan (Agrochemistry and Soil Science), 207 pp

Mashali AM (1995) Integrated soil management for sustainable use of salt-affected soils and network activities. Proceedings of the international workshop on integrated soil management for sustainable use of salt-affected soils. 6–10 November 1995. Bureau of Soils and Water Management, Manila, Philippines, pp 55–75

Massoud FI (1974) Salinity and alkalinity. In: a world assessment of soil degradation. An international program of soil conservation. Report of an expert consultation on soil degradation, FAO, UNEP, Rome, Italy, pp 16–17

Massoud FI (1977) Basic principles for prognosis and monitoring of salinity and sodicity. Proceedings of the international conference on managing saline waters for irrigation. 16–20 Aug 1976, Texas Tech University, Lubbock, Texas, USA, pp 432–454

Massoud FI (1981) Salt affected soils at a global scale for control. FAO Land and Water Development Division Technical Paper, Rome, Italy, 21pp

Metternicht GI, Zinck JA (2003) Remote sensing of soil salinity: potentials and constraints. Remote Sens Environ 85(1):1–20

Oldeman LR, Hakkeling RTA, Sombroek WG (1991) World map of the status of human-induced soil degradation. An explanatory note. Second revised edition. International Soil Reference and Information Center (ISRIC), Wageningen, 35 pp

Pazira E (1999) Land reclamation research on soil physico-chemical improvement by salt leaching in southwest part of Iran. IERI, Karaj

Ponnamperuma FN (1984) Role of cultivar tolerance in increasing rice production on saline-lands. In: Staple RC, Toenniessen GH (eds) Salinity tolerance in plants: strategies for crop improvement. Wiley, New York, pp 255–271

Postel S (1989) Water for agriculture: facing the limits. Worldwatch paper 93. Worldwatch Institute, Washington DC, USA, 54 pp

Qadir M, Quillerou E, Nangia V, Murtaza G, Singh M, Thomas RJ, Drechsel P, Noble AD (2014) Economics of salt-induced land degradation and restoration. Nat Res Forum 38(4):282–295

Shahid SA (2013) Developments in salinity assessment, modeling, mapping, and monitoring from regional to submicroscopic scales. In: Shahid SA, Abdelfattah MA, Taha FK (eds) Developments in soil salinity assessment and reclamation – innovative thinking and use of marginal soil and water resources in irrigated agriculture. Springer, Dordrecht\Heidelberg\New York\London, pp 3–43

Shahid SA, Omar SAS, Grealish G, King P, El-Gawad MA, Al-Mesabahi K (1998) Salinization as an early warning of land degradation in Kuwait. Probl Desert Dev 5:8–12

Shahid SA, Abo-Rezq H, Omar SAS (2002) Mapping soil salinity through a reconnaissance soil survey of Kuwait and geographic information system. Annual research report, Kuwait Institute for Scientific Research, Kuwait, KSR 6682, pp 56–59

Shahid SA, Abdelfattah MA, Omar SAS, Harahsheh H, Othman Y, Mahmoudi H (2010) Mapping and monitoring of soil salinization – remote sensing, GIS, modeling, electromagnetic induction and conventional methods – case studies. In: Ahmad M, Al-Rawahy SA (eds) Proceedings of the international conference on soil salinization and groundwater salinization in arid regions, vol 1. Sultan Qaboos University, Muscat, pp 59–97

Shahid SA, Abdelfattah MA, Wilson MA, Kelley JA, Chiaretti JV (2014) United Arab Emirates keys to soil taxonomy. Springer, Dordrecht/Heidelberg/New York/London, 108 pp

Soil Survey Staff (2014) Keys to soil taxonomy 12th ed. US Department of Agriculture, Natural Resources Conservation Service, US Government Printing Office, Washington DC, 360 pp

Szabolcs I (1974) Salt-affected soil in Europe. Martinus Nijhoff, The Hague, 63 pp

Szabolcs I (1989) Salt-affected soils. CRC Press, Boca Raton, 274 pp

Tanji KK (1990) Nature and extent of agricultural salinity. In: Tanji KK (ed) Agricultural salinity assessment and management. ASCE manuals and reports on engineering practice no 71, ASCE New York, USA, pp 1–17

UNEP (1992) Proceedings of the Ad-hoc expert group meeting to discuss global soil database and appraisals of GLASOD/SOTER, 24–28 February 1992, Nairobi, Kenya, 39 pp

Wiley GR (1953) Prehistoric settlement patterns in the Viru Valley, Peru. Smithsonian Institute, Bureau of American Ethnology, Bulletin 155 Washington DC, USA XXII+454 pp (+60 plates)

Chapter 3
Salinity and Sodicity Adaptation and Mitigation Options

Shabbir A. Shahid, Mohammad Zaman, and Lee Heng

Abstract Soil salinity and sodicity are twin constraints to agriculture production in many countries causing significant losses of crop production and land degradation. Once the salinity and sodicity problems are properly diagnosed, an integrated soil reclamation program may be formulated including combination of physical, chemical, hydrological and biological methods to rectify the twin problems. A combination of adaptation and mitigation technologies are to be adopted, for example adaptation allows the continued use of salt-affected soils by adjusting in response to the degree by which salinity and sodicity development has affected the soil, whereas, in contrast, mitigation refers to the technologies which are adopted to stop salinization to occur. It should be remembered that there is no single universal mitigation technology suitable for all soils, however, diagnostic based recommendations work satisfactorily for a specific site or location. Prior to setting up soil reclamation plan it is essential to review the available resources (farmer budget, availability and quality of water) and the objectives of reclamation and the reclamation plan established suiting the specific farmer needs. In this chapter, various soil reclamation methods such as; physical-leveling, subsoiling, mixing sand, seed bed preparation and salts scrapping); chemical (use of gypsum based on gypsum requirement, sulfur, acids etc.), hydrological-selection of suitable irrigation system-drip, sprinkler, bubbler, furrow, using the concept of leaching requiring/fraction to manage rootzone salinity, flushing, drainage, blending of water etc.; biological (use of organic amendments, green manuring, farm yard manures and selection of salt-tolerant crops) have be described. In addition, various methods of screening crops against salinity including hydroponics, field screening and serial biological concentration approach are described. Climate Smart Agriculture practices, integrated soil fertility management using **4 R** nutrient stewardship are concisely reported. Procedures of salt-harvesting from saline lands and deep deposits and their commercial exploitation in industries are also introduced.

Keywords Salinity · Sodicity · Adaptation · Mitigation · Soil reclamation · Leaching requirement · Salt harvesting

M. Zaman et al., *Guideline for Salinity Assessment, Mitigation and Adaptation Using Nuclear and Related Techniques*, https://doi.org/10.1007/978-3-319-96190-3_3

1 Introduction

Soils affected by salinity and sodicity are not confined to just arid and semi-arid regions, where rainfall is insufficient to leach salts from the soil. Saline and sodic soils have been recorded in a wide range of environments under many different hydrological and physiographic conditions. Such a wide distribution tells us that '***no single adaptation or mitigation option or technique will be applicable to all land areas***'. However, diagnostics-based, site-specific recommendations can be viable options. Most importantly, they suggest the formulation of an integrated reclamation and management plan, one which is based on the major constraints facing us – available resources and variable environmental conditions.

First, one needs an integrated system of soil reclamation and management techniques, and these can generally be grouped into four adaptation or mitigation approaches to deal with salt-affected soils. The four approaches are: (i) Hydrological, (ii) Physical, (iii) Chemical, and (iv) Biological, though under unique environmental conditions salt-affected soils may be used for other purposes.

2 Mitigation and Adaptation Options

Mitigation and adaptation are terms commonly used in the context of climate change and many scientists believe that mitigation is primarily concerned with emission of greenhouse gases (GHGs), while adaptation deals with water and agriculture. However, mitigation and adaptation also apply to how mankind must deal with salt-affected soils. In this chapter, we have defined both terms in the context of salt-affected soils and their reclamation and management.

Adaptation allows the continued use of salt-affected soils by making adjustments in response to the degree by which salinity and sodicity development has affected the soil. In contrast, mitigation refers to the technologies which are adopted to stop salinization to occur.

3 Diagnostics of the Soil Salinity Problem

Soil salinity in agricultural fields is increasing worldwide, mainly due to poor farm management practices and the increasing demand for intensification of agriculture for the short-term benefits of increased food production. This intensification ignores the long-term consequences on 'other services' provided by the soil. It is, therefore, very important to understand the salinity hazard, both spatially and temporally at the regional and national farm levels.

In agricultural fields, an effective measurement of salinity will identify the location and extent of root-zone salinity and ensure that root-zone salinity is kept below the threshold level for each crop. Soil salinity is dynamic and has a wide variation vertically, horizontally, and temporally. Many consider soil salinity a uniform feature in a soil profile. However, Shahid et al. (2009) showed that, for the saline-sodic soils of Pakistan, salinity was a layered feature as one moved down the profile.

At the regional (Middle East) and national (Kuwait, United Arab Emirates) levels, a salinity mapping program (Shahid et al. 2010) helped policy makers in taking necessary and timely actions to tackle the issue of increased soil salinity. Similar programs will help to avoid a further spread of soil salinity to new regions; and they, if prove successful, will prevent negative impacts on national economics through degrading a nation's soil resources.

4 Integrated Soil Reclamation Program (ISRP)

Salt-affected soils are distributed across a wide range of hydrological and physiographical conditions, soil types, rainfall and irrigation regimes, as well as different socioeconomic settings. This diversity makes one realize that there will be no single technique of soil reclamation applicable to all areas. The exploitation of these soils for agriculture will require an integrated reclamation and management plan based on a comprehensive investigation of soil characteristics, including water monitoring (rainfall, irrigation and soil water-table), a survey of crops and local conditions, including climate, the economic, social, political, and cultural environment, as well as the existing farming systems. Fortunately, several approaches can be combined into an integrated system of soil reclamation and management (Shahid and Alshankiti 2013).

4.1 Objectives of Salinity Reclamation

The main objectives of reclamation are to:

- Improve soil health for better crop production
- Bring abandoned farms back to cultivation
- Increase the crop yield per unit of land area
- Improve food security within national boundaries
- Enhance water and fertilizer use efficiencies
- Optimize cost of crop production per unit area, and
- Improve the livelihood of the farmers

4.2 Prerequisite for Soil Reclamation

A soil reclamation plan can only be implemented if certain prerequisites are fulfilled. Some of these are essential for efficient, effective and long-term reclamation of salt-affected soils, as listed below.

- The farmer is convinced and ready to initiate soil reclamation at his farm
- The farmer has sufficient financial resources to implement the plan
- It is essential to have land leveling by laser before the initiation of any reclamation plan; this will help to ensure uniform water distribution and effective leaching of salts
- Additionally, a supply of good quality water is required
- Good subsurface drainage of the soil to be reclaimed is essential
- There must be a plan to handle drainage water safely, without compromising the environment

Sustainable agriculture with salt-affected lands is most likely to be achieved through integrated soil reclamation program (ISRP) and natural resource management (NRM). These are approaches where the long-term condition of the resource is built in as a core consideration. Thus, crop production on salt-affected lands can only be successful if the soil is dealt with in a holistic manner, i.e. in an '*integrated approach*' which includes all aspects of soil, water, plants and climatic conditions. There is, unfortunately, a misconception about ***Biosaline Agriculture***, that this is a complete solution for using salt-affected lands and saline and saline-sodic waters. Rather, biosaline agriculture is just one of the components of ISRP, which also includes physical, chemical, hydrological and biological methods (Box. 3.1).

Box 3.1 – Integrated Soil Reclamation Program (ISRP)

It should be noted that no single approach can deliver a complete solution to fix/reclaim soil salinity problem. This means that we need to take holistic approach by using a combination of mitigation approaches, which could be site-specific, and should only be used in other areas where similar soils and environmental conditions exist. International Center for Biosaline Agriculture (ICBA), Dubai, UAE has sufficient expertise in the diagnostics of the problem, developing an integrated reclamation strategy and implementation of this strategy to transform marginal soils to good quality for crop production, using methods, such as ***physical*** (leveling, salt scraping, tillage, subsoiling and sanding); ***chemical*** (use of soil amendments such as elemental S, acids, gypsum, etc. based on gypsum requirements to rectify soil sodicity problems and to improve soil health); ***hydrological*** (irrigation systems: Surface, flood, basin, drip, sprinkler, subsurface irrigation, etc., and leaching and drainage), and ***biological*** (biosaline agriculture: Salt tolerant crops, and a serial biological concentration approach). (Adapted from Shahid and Rahman (2011) and Shahid et al. (2011))

It should be further noted that the exploitation of salt tolerant crops and saline waters, i.e. '***Biosaline Agriculture***', without adopting other components of integrated soil reclamation, will ultimately degrade the soils further. These degraded soils will likely be unable to provide essential soil services, as such, but not limited to, agricultural production.

As discussed briefly above, recently established worldwide strategies for soil reclamation can be grouped into the following approaches which cover almost all aspects of soil reclamation and management.

- Physical
- Chemical
- Hydrological
- Biological
- Alternative land uses

4.3 Physical Methods of Soil Reclamation

There are several physical methods of soil reclamation, though not all of these methods are required for a given situation. Site-specific diagnostics can allow one to select the most suitable method(s). The most commonly used physical methods of soil reclamation are listed below.

- Leveling
- Subsoiling – deep plowing and deep ripping
- Mixing sand with heavy textured soil – sanding
- Seed bed shaping to reduce salinity effects – tillage practices
- Physical removal of the surface salt crust – scraping salts

4.3.1 Leveling

Leveling of salt-affected lands prior to the implementation of a reclamation program is essential. This allows for a uniform water distribution, leading to effective leaching of salts. Unfortunately, the farmer usually accomplishes this task by plowing the field, followed by the use of a conventional planking tool. This practice usually leaves the land uneven, which means that when water is applied to the field in order to leach the salts, the water puddles in the depressions and heterogeneous conditions are formed. There exists a modern tool '*laser land leveling*' to level the land in highly effective manner. It is, thus, suggested that each farmer should contact the extension services department in order to access this modern tool, which will facilitate an effective initiation of the farmer's soil reclamation program. The leveling process, however, may compact the soil due to the use of heavy machinery. If this occurs, the leveling process should be followed by subsoiling or chiseling.

4.3.2 Subsoiling

The soils affected by sodicity are usually underlain by dense clay-sodic layer(s). These dense layers are created by the dispersion of clay particles in the highly sodic water. The dispersed clay particles move to the subsurface of the soil where they are lodged on the surfaces of the conducting soil pores, thus, blocking the pores and preventing further water movement. It is particularly important, then, to disrupt the dense layers deep in the soil in order to enhance permeability. This is especially important for reclaiming sodic soils after the addition of amendments such as gypsum, followed by watering the field. The addition of gypsum will enhance the removal of exchangeable sodium (which has already been exchanged by Ca^{2+}) into lower layers of the soil prior to finally moving into the main drainage system. In addition to dense sodic layers, the soils may be underlain by a plow layer or by other hard pans occurring during soil formation. These hard pans must be disrupted and broken in order to enhance drainage capacity, and to facilitate the soil reclamation process.

4.3.3 Sanding

If the soil to be reclaimed has heavy texture (i.e., a clay soil), the mixing of sand in an appropriate quantity can change the soil texture permanently; the soil becomes more permeable and is easier to reclaim. This practice also provides a favorable environment for plant growth compared to the original soil prior to sanding. Changing the soil texture is a difficult and costly task, though where sand is readily available, such as in a desert, this practice can be accomplished more easily.

A clay soil is considered to exist in an area when it has the percentages of primary soil particles, as: 10% sand, 20% silt, and 70% clay (Textural class: Clay). The clay soil is mixed with a known quantity of sand to develop the following percentages of soil particles: 60% sand, 15% silt, and 25% clay (Textural class: Sandy clay loam).

In this way, the original soil texture (Clay) is significantly changed to another soil texture (Sandy clay loam). Both soil textural classes can be located on the USDA soil textural triangle (Fig. 3.1).

The newly established texture, a sandy clay loam, has improved soil physical properties, e.g. an increased drainage capacity and infiltration rate. This leads to an enhanced soil reclamation process and results in a much better leaching of salts.

4.3.4 Scraping

Salt accumulation in an irrigated field is common. Salts accumulate at certain zones based on the irrigation system in use and soil bed shape. The readers are referred to Chap. 4 of this book for more information in this respect. To clarify the scraping practice in order to remove salts, a furrow irrigation system is selected (Plate 3.1);

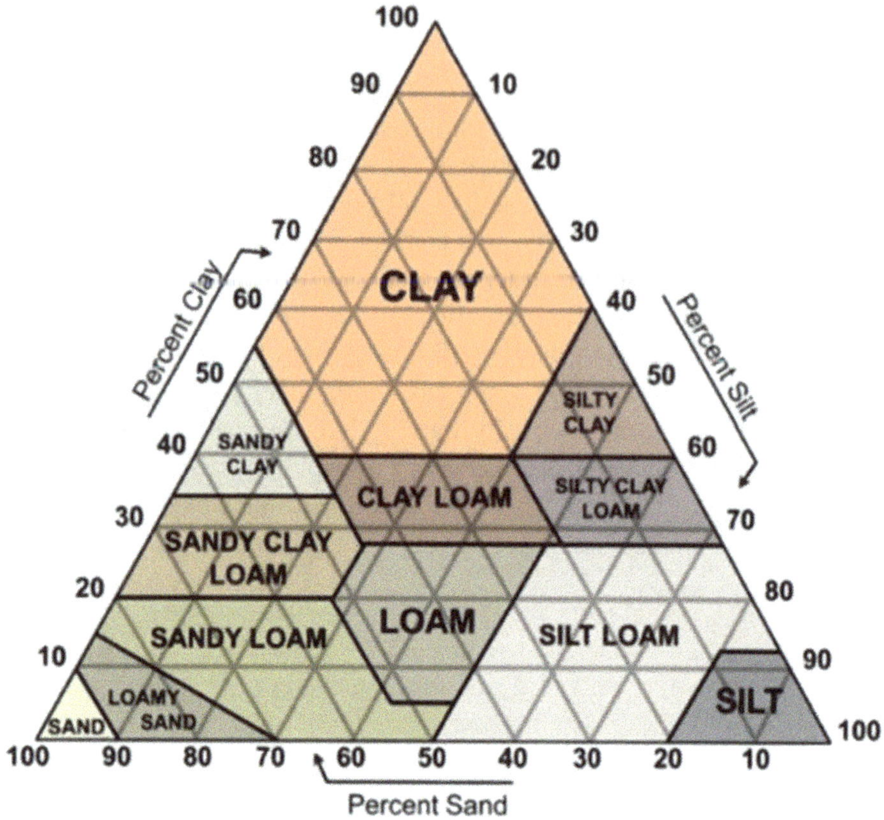

Fig. 3.1 USDA soil textural classes

Plate 3.1 Salts buildup in furrow irrigation system

both furrows are irrigated and the zone of salt accumulation appears in the center of the ridge. The salt crusts which accumulate at the surface, through capillary rise and subsequent evaporation, can be removed manually or through mechanical means. The mechanical removal is the simplest and the most economical way of reclaiming saline soils, if the area is small. This practice minimizes the salts temporarily; they will be accumulated again if there is a continuous feed of saline water to the soil surface. Therefore, the scraping method is considered a temporary solution.

4.3.5 Seed Bed Preparation – Tillage

Tillage can improve the soil's physical condition, and can bring a saline layer to the surface. However, it can also create a plow layer through the continuous use of the plow. Care must, thus, be exercised in using tillage. By manipulating the soil surface into different shapes and the selection of a specific irrigation system, a zone of low salinity can be achieved. It is well recognized that salts tend to accumulate on the ridges away from the wet zone when furrow irrigation is adopted. Placing the seeds on the off-center slope (i.e., shoulder) of the single row will position the seed in a location with a minimum salinity and an optimum moisture condition. Under high salinity conditions, the alternate row should be left un-irrigated. This will ensure maximum accumulation of salts in the un-irrigated area, thereby leaving the irrigated furrows free of salts and fit for planting seeds. The readers are referred to Chap. 4 for a detailed account.

4.4 Chemical Methods of Soil Reclamation

It should be kept in mind that chemical methods of reclamation are commonly used to reclaim sodic or saline-sodic soils. Saline soils are unable to be reclaimed by chemical methods. Chemical reclamation includes the use of gypsum, elemental sulfur and acids (hydrochloric and sulfuric acids), and the methods used are based on the diagnostics of the problem. Sodic soil can be recognized through visual assessment in the field, or through analyzing soils in the laboratory for exchangeable sodium percentage (ESP). A soil with ESP > 15 is classed as a sodic soil (USSL Staff 1954). At this threshold ESP value, the soil will show effects on both soil physical properties (structural damage) and also on plant growth. In such soils, the objective is to bring the soil ESP below the threshold value. This can be achieved by adding suitable amendments to increase the concentration of calcium ions (Ca^{2+}) in the soil.

4.4.1 Use of Gypsum to Reclaim Sodic Soil

The most suitable method is to replace exchangeable sodium with calcium, and the subsequent use of organic matter to bind the soil and improve its structure. Gypsum ($CaSO_4.2H_2O$) and lime (CaO) can both add calcium and, thus, can overcome dispersion as the calcium causes the inter particle forces to more readily hold the particles together. The calcium causes particles to form clusters (flocculates), forming a very clear puddle of water. Gypsum usually gives an immediate response as it dissolves (although it has low solubility) in water, though it leaches sooner than lime. Amendments are those materials which supply Ca^{2+} for the replacement of exchangeable sodium and furnish calcium indirectly by dissolving calcite ($CaCO_3$), which is naturally occurring in many arid zone soils. Gypsum is reported to reduce the levels of exchangeable sodium in the soil. It also improves both soil tilth and drainage, and achieves better crop production.

The addition of gypsum to a soil changes the soil chemistry in two ways: i) by increasing the amount of salt which is in solution, thereby avoiding the swelling and dispersion of the clay component. This is a short term effect which occurs as the gypsum dissolves, and ii) the calcium from the gypsum replaces the exchangeable sodium which was adsorbed onto the clay at specific sites. This process changes sodic clay to a calcium-clay. The displaced sodium is then leached into lower soil zones, below the plant root-zone. Mined gypsum (less than 2 mm particle size) of commercial grade (~ 70% purity) is commonly used for the reclamation of sodic soils.

The following reaction occurs when gypsum is added to a Na-clay soil.

$$\underset{\text{Sodic soil}}{2Na^{+}-\text{Clay}}+\underset{\text{Gypsum}}{CaSO_4.2H_2O}\rightarrow\underset{\text{Normal soil}}{Ca^{2+}-\text{Clay}}+\underset{\text{Salts leach out}}{Na_2SO_4^{-}+2H_2O}$$

4.4.1.1 Determination of Gypsum Requirement

Gypsum requirement (GR) is the quantity of gypsum needed per acre or per hectare of soil to lower the exchangeable sodium percentage (ESP) of the soil to a desired level.

4.4.1.2 How to Determine the Weight of One-Hectare Soil?

Laboratory methods for measuring gypsum requirement of a soil are based on meq 100 g^{-1} of dry soil. Thus, one needs to convert the GR in meq 100 g^{-1} to metric tons (weight) of soil present in a hectare at either 15 or 30 cm depth. The area covered by a hectare of land is 10,000 m^2 (100 m × 100 m). The weight of one-hectare soil of varying depths (15 or 30 cm) can be determined by the following procedure.

- First, determine the bulk density of soil by taking a cylindrical core (Plate 3.2) of soil with a known volume (e.g., a diameter of 8 cm, and a height of 5 cm).
- Remove the soil from the core and oven-dry it at 105 °C.
- Weigh the oven-dried soil (g).
- Determine the bulk density of the soil by using the standard calculation, as below:

$$\text{Bulk density} = \text{Mass of the soil(g)} \div \text{Volume of bulk soil collected } \left(\text{cm}^3\right)$$

Plate 3.2 Collection of a soil core to measure bulk density of the soil, and cleaning the core with a saw

Bulk volume is, thus, defined as the volume of soil occupied both by mineral matter of the soil and the space between the mineral matter particles.

Example

- Standard core size (diameter 8 cm; height 5 cm)
- Core volume ($\pi r^2 h$)
  where $\pi = 3.143$, r = radius of the cylindrical core in cm; h = height of the core in cm.
  $= 3.143 \times 4 \times 4 \times 5 = 251.44\ cm^3$
- Weight of the oven-dry sandy soil from the core = 402.4 g
- Bulk density of sandy soil = mass per unit bulk volume = $402.4 \div 251.44 = 1.60$ gram per cubic centimeter ($g\ cm^{-3}$)
- Volume of one-hectare to a 30 cm depth (length × width × depth)
  $10{,}000\ cm \times 10{,}000\ cm \times 30\ cm = 3{,}000{,}000{,}000\ cm^3$
- Weight of 1 cm^3 of soil (1.60 g)
- Weight of 3,000,000,000 cm^3 soil = 4,800,000,000 g or 4,800,000 kg
- Thus, there are 4.8 million kilograms in a one-hectare volume of soil to the 30 cm depth, and 2.4 million kilograms to the 15 cm depth. In order to know if the soil needs the application of gypsum for reclamation, it is necessary to first diagnose the problem through field investigation (Plate 3.3). Once the soil sodicity is diagnosed, soil samples must be collected at different soil depths in order to assess the average level of sodicity. Then, the gypsum requirement can be calculated for sodicity reclamation.

4.4.1.3 Conversion of Gypsum Requirement – Lab Results to Field Application

Normally, the gypsum requirement (GR), determined through lab procedure, is in milli equivalent per 100 g of soil (meq 100 g^{-1}), which needs to be calculated as

Plate 3.3 On-site diagnosis of a saline-sodic soil in Pakistan, and sharing the experience with farmers

metric tons per hectare for 15 or 30 cm soil depth for the purpose of field application. The factor for conversion of the gypsum requirement values from meq 100 g^{-1} to tons per hectare is explained here. Based on the bulk density (1.60 g cm^{-3}) of the soil in the above example (UAE sandy soil average bulk density), 4.8 million kilograms weight of soil was determined to be present in one hectare area of soil to a 30 cm depth. Using these figures, a factor of 2.066 has been determined to convert the GR of 1 meq 100 g^{-1} soil to metric tons per hectare of soil to the 15 cm depth, and a factor of 4.132 to a 30 cm depth.

The USSL Staff (1954) has reported a factor of 0.86 to convert gypsum requirement (GR) from 1 meq 100 g^{-1} soil to tons per acre at a 15 cm depth. Since one hectare is equal to 2.471 acres, a factor of 2.125 is used in USDA Handbook 60 to convert GR from 1 meq 100 g^{-1} to tons per hectare at the 15 cm depth, and a factor of 4.250 for tons per hectare to a 30 cm depth.

4.4.1.4 A Comparison of Gypsum Requirement Between USSL Staff (1954) and Sandy Soils of United Arab Emirates

Assume we have analyzed five soils (A, B, C, D and E) for their gypsum requirement. These five soils have shown a GR of 1, 2, 3, 4 and 5 meq 100 g^{-1}, respectively. Calculate the GR of these five soils in terms of tons per hectare to the 15 and 30 cm depths using the above conversion factors. The calculated GR for the five soils is presented in Table 3.1.

It should be noted that the use of the GR conversion factor from the USSL Staff (1954) publication will slightly overestimate the GR (e.g, 4.250 tons per hectare versus 4.132 tons per hectare to the 30 cm depth for soil A). Therefore, it is advisable that countries where the bulk density of the soils is very different to that of most US soils, a new factor should be determined for each of the local soils.

Table 3.1 Gypsum requirement of five soils determined in lab and calculated in tons per hectare using conversion factors

	Soil gypsum requirement				
		Metric tons hectare^{-1} (USSL Staff 1954)		Metric tons hectare^{-1} (UAE soils)	
Soil	meq 100 g^{-1} (Lab results)	15 cm depth	30 cm depth	15 cm depth	30 cm depth
A	1	2.125	4.250	2.066	4.132
B	2	4.250	8.500	4.132	8.264
C	3	6.375	12.750	6.198	12.396
D	4	8.500	17.000	8.264	16.528
E	5	10.625	21.250	10.330	20.660

4.4.1.5 Gypsum Requirement

Method 1 (Schoonover 1952)

The procedure is described as follows.

- 5 g soil + 100 ml gypsum saturated solution (GSS) → 5 m of shaking in the mechanical shaker → filter and titrate for Ca^{2+} + Mg^{2+} (in meq l^{-1})
- GR meq 100 g^{-1} = (Ca^{2+} + Mg^{2+} meq l^{-1} in GSS) – (Ca^{2+} + Mg^{2+} meq l^{-1} in filtrate) x 2
- Note that a factor of 2 is used to convert GR from meq l^{-1} to meq 100 g^{-1} of soil; derivation of this factor of 2 is explained below.
- Assume the GR is determined as x meq l^{-1}, then
  - 1000 ml of soil solution requires the GR = x meq
  - 100 ml of soil solution requires GR = (x/1000) × 100 = x/10 meq
  - Or, 5 g of soil requires GR = x/10 meq
  - 100 g of soil requires GR = x/10 × 1/5 × 100 = 2x (Note: Factor of 2 only works when 5 g soil is used in 100 ml GSS, different factors exist for different soil quanity).
- Assuming that the GR of a sodic soil is 1 meq 100 g^{-1} of soil, we need to determine how many metric tons of gypsum should be added to soil per hectare for each of the 15 and 30 cm depths.
- Equivalent weight of gypsum ($CaSO_4.2H_2O$) = 86.09 g
  - 1 equivalent of Na^+ will require 86.06 g of gypsum.
  - 1 meq of Na^+ will require 0.08606 g of gypsum.
  - Weight of one-hectare soil is 4,800,000 kilograms to the 30 cm depth, when bulk density is 1.60 g per cm^3.
  - The GR is, thus, 4.132 metric tons per hectare to the 30 cm depth or 2.066 tons per hectare to the 15 cm depth.
  - 1 hectare = 2.471 acres

In the above procedure, an equivalent amount of soluble CO_3^{2-} and HCO_3^- are also precipitated; therefore, this method accounts for exchangeable sodium and also for soluble CO_3^{2-} and HCO_3^- in the soil.

The calculation, however, is based on 100% pure gypsum. Commercial grade gypsum purity is about 70%. Therefore, a factor based on purity must be calculated in order to correct the GR requirement for the commercial grade gypsum.

Method 2 (USSL Staff 1954, Modified by Shahid and Muhammed 1980)
In this method, the gypsum requirement is calculated based on the exchangeable sodium and cation exchange capacity values taken from the laboratory analyses.

Example 3.1
A soil was analyzed in an accredited laboratory and following results were obtained.

- Exchangeable sodium (ES) = 4 meq 100 g^{-1}
- Cation exchange capacity (CEC) = 10 meq 100 g^{-1}
- Exchangeable sodium percentage (ESP) = (ES/CEC) × 100
- Thus, ESP will be 40. In order to reduce ESP from 40 to 15 (threshold value), one would need to add gypsum equivalent to 2.5 meq 100 g^{-1} of exchangeable Na^+. Thus, as above:
- 1 equivalent of exchangeable Na^+ 100 g^{-1} soil will require 86.09 g of gypsum.
- 1 meq of exchangeable Na^+ 100 g^{-1} soil will require 0.08609 g of gypsum.

Calculate the gypsum requirement (metric tons per hectare for the 30 cm soil depth), keeping in mind that the weight of one-hectare dry soil is 4.8 million kilograms.

From Example 3.1, it was determined that soil with a bulk density of 1.60 g per cm^3, a factor of 4.132 can be used to convert the GR from meq per 100 grams to metric tons per hectare for the 30 cm depth. Therefore, a GR based on the 2.5 meq per 100 grams will be equal to 10.330 metric tons of gypsum per hectare. However, the conversion factor for commercial grade gypsum should be applied to the 10.330 metric tons value.

Example 3.2
A soil was analyzed in an accredited laboratory and following results were obtained.

- Exchangeable sodium (ES) = 2 meq 100 g^{-1}
- Cation exchange Capacity (CEC) = 5 meq 100 g^{-1}
- Exchangeable sodium percentage (ESP) = (ES/CEC) × 100
- Thus, ESP will be 40. In order to reduce ESP from 40 to 15 (threshold value), we need to add an amount of gypsum which is equivalent to 1.25 meq 100 g^{-1} of exchangeable Na^+
- 1 equivalent of exchangeable Na^+ 100 g^{-1} soil will require 86.09 g of gypsum
- 1 meq of exchangeable Na^+ 100 g^{-1} soil will require 0.08609 g of gypsum

Calculate gypsum requirement (metric tons per hectare for the 30 cm soil depth), keeping in mind that the weight of one-hectare dry soil is 4.8 million kilograms.

From the example above, it was determined that for a soil with a bulk density of 1.60 g per cm^3, a factor of 4.132 can be used to calculate (convert) the GR from meq per 100 grams to metric tons per hectare for the 30 cm depth. Therefore, the GR based on adding the 1.25 meq per100 grams of gypsum will be equal to 5.165 metric tons per hectare.

From above examples, we can conclude that even if two soils have same Exchangeable Sodium Percentage (ESP), the gypsum requirement can be significantly different.

It should be noted that in this procedure of determining GR, an equivalent amount of soluble CO_3^{2-} and HCO_3^- (meq 100 g^{-1}) must be added (taken into account) in order to properly calculate the total gypsum requirement. This is because a gypsum equivalent to CO_3^{2-} and HCO_3^- will be precipitated. This modification to USSL Staff (1954) method was made by Shahid and Muhammed (1980).

4.4.2 Use of Acids to Reclaim Calcareous-Sodic Soils

It should be noted that the use of acids is recommended for sodic soils which are also calcareous. The acids react with the calcium carbonates present in soil to mobilize calcium, which ultimately replaces exchangeable sodium in the soil, thereby reducing the ESP. The objectives of acids application are to:

- Mobilize calcium from calcium carbonates
- Replace exchangeable sodium with calcium
- Bring about a reduction in soil pH, thereby enhancing nutrient uptake, and
- Improve soil health in order to obtain a better crop production

Both sulfuric and hydrochloric acids react rapidly with soil lime, since they do not have to go through an oxidation process. However, they are highly corrosive, and dangerous to handle. Specialized equipment has recently become available to safely apply acid onto field soil, usually with irrigation water. The reaction of applied acids with naturally occurring soil $CaCO_3$ and exchangeable Na^+ is shown below.

$$CaCO_3 + H_2SO_4 \rightarrow CaSO_4 + CO_2 + H_2O$$

$$\underset{\text{Sodic soil}}{2Na^+ - Clay} + CaSO_4 \rightarrow \underset{\text{Normal soil}}{Ca^{2+} - Clay} + \underset{\text{Salts leach out}}{Na_2SO_4 \downarrow}$$

$$CaCO_3 + 2HCl \rightarrow CaCl_2 + CO_2 + H_2O$$

$$\underset{\text{Sodic soil}}{2Na^+ - Clay} + CaCl_2 \rightarrow \underset{\text{Normal soil}}{Ca^{2+} - Clay} + \underset{\text{Salts leach out}}{2NaCl \downarrow}$$

4.4.3 Use of Elemental Sulfur to Reclaim Calcareous-Sodic Soils

Elemental sulfur can also be used to reclaim calcareous-sodic soils with one condition – the sulfur must be completely oxidized. This occurs through biological

Table 3.2 Equivalent amount of various amendments for supplying Ca in terms of pure gypsum

Amendment	Ton(s) equivalent to 1 ton of 100% gypsum[a]
Gypsum ($CaSO_4.2H_2O$)	1.00
Calcium chloride ($CaCl_2.2H_2O$)	0.86
Calcium nitrate [$Ca(NO_3)_2.2H_2O$]	1.06
Press-mud (lime-sulfur, 9% Ca + 24% S)	0.78
Sulfuric acid (H_2SO_4)	0.61
Iron (Ferrous) sulfate ($FeSO_4.7H_2O$)	1.62
Ferric sulfate ([$Fe_2(SO_4)_3.9H_2O$]	1.09
Aluminum sulfate [$Al_2(SO_4)_3.18H_2O$]	1.29
Sulfur (S)[b]	0.19
Pyrites (FeS_2, 30% S)[b]	0.63
Limestone ($CaCO_3$)	0.58

Adapted from USSL Staff 1954

[a]The quantities are based on the use of 100 % pure materials. If the material is impure, necessary corrections must be made. For example, if the gypsum is a 70 % agricultural grade, the equivalent quantity which must be applied will be 1.43 tons

[b]100% oxidation is assumed though, in practice, it does not happen

Partial source: Ayers and Westcot (1985)

oxidation of sulfur by *Thiobacillus thiooxidans*, although in sodic soils sulfur oxidation is a very slow process. The complete oxidation of sulfur results in formation of sulfuric acid.

$$2S + 3O_2 + 2H_2O \rightarrow 2H_2SO_4$$

The H_2SO_4 formed through biological oxidation of sulfur reacts rapidly with lime (as shown above) and proceeds to reclaim the sodic soils.

There is a significant financial outlay required from farmers when using chemical amendments in soil reclamation. It is, therefore, recommended that the benefit of any amendment should be tested first in field trials with regard to its cost, safety in use, and effectiveness in improving (reducing) soil sodicity and increasing crop production. The theoretical amounts of various amendments to supply an amount of Ca equivalent to 1 ton of gypsum are shown in Table 3.2.

A good example is seen from an experiment with vertisol from India; Sharma and Gupta (1986) observed a similar change in ESP (Table 3.3) due to the application of either gypsum or H_2SO_4. However, they reported that there was a low hydraulic conductivity and higher amounts of water dispersible clay in the case of H_2SO_4 application. This was caused by the Ca^{2+} being a stronger flocculent than the H^+ ion of Sulfuric acid.

Table 3.3 Effect of different amendments (applied @ 100% gypsum requirement) on physical and chemical properties of a sodic vertisol (Sharma and Gupta 1986)

Amendment	Soil characteristics				
	$pH_{1:2}$	EC ($dS\ m^{-1}$)	ESP	HC_{sat}^{a} ($mm\ hr^{-1}$)	Water dispersible clay (%)
Control	8.8	9.80	65	0.06	37.2
Gypsum	7.9	0.72	14	4.77	8.0
Pyrites	8.0	0.31	20	1.64	32.4
H_2SO_4	7.5	0.18	14	2.98	30.4
$Al_2(SO_4)_3$	7.6	0.27	8	4.49	8.6
$FeSO_4$	7.9	0.85	21	1.59	33.7

[a]Saturated hydraulic conductivity

4.5 *Hydrological Methods of Soil Reclamation*

Hydrological methods generally include irrigation, leaching and flushing, and the drainage of the leached water. Blending of water to reduce its salinity and sodicity, and recycling the water can also be included under this topic.

The objectives of soil reclamation through hydrological methods are:

- Efficient use of irrigation water
- Leaching of salts into lower soil zones, below the root-zone
- Improvement of water quality
- Improvement of a waterlogged condition through drainage

In irrigated agriculture, the salts in soil can be removed in two ways:

- Leaching of salts into a soil zone below the root-zone and subsequent drainage of the leached water to a safe place, and
- Surface flushing of dissolved salts

The salt content within the root-zone is likely to be increased if the net downward movement of salts is less than the salt input from irrigation water. Therefore, the soil salt balance must be kept under control. Control of the salt balance is, thus, a function of irrigation water quality, the quantity of dissolved salts in the water, and the success of the soil drainage system.

4.5.1 Leaching

Soils rich in soluble salts can be reclaimed through dissolving of these salts and their successful leaching. This can be accomplished through flooding or ponding of water at the surface for saline soils. In general, the depth of soil leached is roughly equal to the depth of water infiltrated during leaching. In order to leach salts from a soil, an understanding about the leaching requirement (LR) concept is important. The LR is the calculated fraction (depth) or quantity of water that must pass through the root-zone in order to maintain the EC of the drainage water at or below a specified level. Some soil scientists are of the opinion that LR should be minimized to prevent raising the level of the groundwater table, and also to reduce the load placed on the drainage system (Mashali 1995). Recently established guidelines for successful and economic leaching methods are described hereunder.

4.5.1.1 Timing of Leaching Irrigation

Timing of leaching does not appear to be critical, provided crop salinity tolerance limit is not exceeded for extended periods of time, or occurs during a critical stage of plant growth. The leaching can even be accomplished at each irrigation event. However, in a soil with a low infiltration rate, and for crops which are sensitive to excess moisture in the root-zone, leaching at each irrigation event may not be possible or advisable. Few important points to consider are listed below.

- Leaching should be done when soil moisture is low and water-table level is deep; and leaching should precede the critical growing stage of the crop plant
- An optimal time for leaching would be during a period with a low evapotranspiration demand, at night, during high humidity and in cooler weather
- Leaching can also be done at the end of the cropping season
- Soil and tissue analysis can help determine both the need and timing of leaching

Sandy soils in a desert environment, such as Gulf Cooperation Council (GCC) countries and other similar environments are well drained. Therefore, leaching of salts by using irrigation water amounts in excess of evapotranspiration can maintain salts in the root-zone to a safe limit. However, one major problem for irrigated agriculture under hot desert conditions is the high amount of drainage water which must be managed safely and sustainably, without compromising the environment.

4.5.1.2 Leaching Requirement and Leaching Fraction

Quantity of water that must pass through the root-zone to maintain the EC level at or below a specified level defines the 'leaching requirement'. The 'leaching fraction' is the fraction of irrigation water that passes through the root-zone, into lower soil zones.

4.5.1.3 Leaching Requirement for Surface Irrigation

In order to determine the leaching requirement, it is essential to have information about two parameters: (i) salinity of the irrigation water to be used (dS m^{-1}), and (ii) crop tolerance to salinity (ECe in dS m^{-1}). Using the equation of Rhoades (1974) and Rhoades and Merrill (1976), the leaching requirement can be calculated as:

$$LR = \frac{ECiw}{(5ECe - ECiw)}$$

Where, ECiw is the salinity of the irrigation water (dS m^{-1}) and ECe is for a given crop's maximum yield potential. LR refers to the minimum leaching requirement that is necessary to control salts within the tolerance limit of the crop, when the crop is grown under an ordinary surface irrigation method.

4.5.1.4 Leaching Requirement for Drip Irrigation System

$$LR = \frac{ECiw}{2(\text{Max EC})}$$

Where, ECiw is the EC of the irrigation water, and a factor of 2 is obtained from ECsw, which is equal to 2ECe.

Knowing the desired leaching requirement (LR) and evapotranspiration (ET) demand of the crop, the net water required for a crop can be calculated (Ayers and Westcot 1985), as below.

$$\text{Net water requirement} = \frac{ET}{(1 - LR)}$$

Where, net water requirement = depth of applied water (mm $year^{-1}$), ET = total annual crop water demand (mm $year^{-1}$), and LR = leaching requirement expressed as a fraction (leaching fraction).

4.5.2 Flushing

Flushing is suitable for saline soils which have surface salt crusts – a common situation in arid and semi-arid areas, and where rainfall is insufficient to leach the salts. This practice flushes the salts from soil surface and the flushed saline water then enters the drainage system, which becomes concentrated with salts. Flushing of surface salts is possible where soils are of a heavy texture and ponding can be accomplished easily. Once the water is ponded for a time which is sufficient to dissolve the salt crust, the ponded water can be flushed from the field, thus removing the surface salts. The following procedure will allow for successful flushing of surface salts.

- A sufficient volume of good quality water is used to dissolve salts from the soil surface
- The soil must possess the ability to allow for surface ponding, e.g. its subsurface must have a heavy texture

- The field must be capable of flushing the dissolved salts. This can be accomplished either by forming breaks in the sides of the field so that the saline water will drain into adjacent channels. Alternatively, the ponded saline water can be removed by siphoning it, using long pipes, into adjacent channels
- There must be ways to safely reuse the drained water, or environmentally safe methods available for its disposal

5 Drainage and Drainage Systems

Drainage is the natural or artificial removal of surface and subsurface water. Land areas which have waterlogged soils, or have a high (shallow) water-table, will require removal of the water if crop production is an objective.

Why drainage?
There must be a strong justification for installing a drainage system, considering the following points.

- Drainage is required to lower water-table
- Drainage is needed to address waterlogging and to bring the land back into crop production
- Drainage is needed to minimize the upward movement of groundwater and to control the buildup of salts due to capillary rise of the groundwater
- Salinity management will be necessary to improve crop production

5.1 Agricultural Drainage Systems

An agricultural soil affected by high water-table requires a drainage system to improve crop production and/or to manage water supplies. There are two types of drainage systems; surface and subsurface.

Depending upon the site conditions, nature of the problem, available resources, different types of drainage systems can be used, these are:

- ***Surface drainage*** – to allow for the runoff of excess water before it enters the soil
- ***Subsurface drainage*** – to control the groundwater table at a lower (safer) depth, by using either open ditches and tile drains or perforated plastic pipes. Methods include passive mole drainage, and also vertical drainage (pumping water) when the deep soil horizons have an adequate hydraulic conductivity.

5.1.1 Surface Drainage – Natural Drainage

This is the cheapest and easy way of draining the water and is possible where the underlying layers are permeable and relief is adequate. However, these ideal conditions do not always exist in saline areas, and a drainage system will always be required there.

5.1.2 Subsurface Drainage

This is the most suited drainage system for irrigated agriculture. It aims at controlling groundwater level, as well as leaching excess salts from the plant root-zone in order to keep the salt balance in soil water below the crop threshold. There are two types of subsurface drainage systems, open ditches and closed drains.

Open drains are deep earth ditches where groundwater flows and is ultimately discharged to a safe place for further use.

Closed drains are pipe drains installed in the field. They collect water and discharge it into a sump whose outlet leads to lagoons or basins.

5.1.3 Tile Drainage System

Tile drainage is a very effective way of controlling water-table and reducing waterlogging in areas where the soil aquifer cannot be pumped. It involves the installation of slotted PVC pipe (or other material) at about 1 meter below the soil surface. Soil water enters the pipe through the slots and is carried to a central well (pit) where it can then be removed, either by pumping or via gravity drainage. Tile drainage can be very expensive.

In order to assure the sustainability of the tile drainage system, it is important to accomplish the following checks on a regular basis.

- Drill test bores at a number of sites on the field
- Monitor water levels in these bore holes on a regular basis
- Check water quality (salinity and sodicity) on a regular basis

5.1.4 Mole Drainage System

In the mole drainage system, subsurface circular channels are developed by use of a mole plow for drainage, functioning like pipes buried in the soil. The success of a mole drainage system depends on the soil properties. Soils with a heavy texture are ideal for mole drainage as they are less vulnerable to collapse. Water continuously enters the mole channels, and the channels usually remain stable for a long time. The mole drainage system is much cheaper than a tile drainage system, and is usually developed on a closely spaced basis, yielding effective drainage. The only drawback

Plate 3.4 Vertical drainage through installing tube well (an example from Pakistan); the poor quality groundwater is used to irrigate salt tolerant plants at the Biosaline Research Station of NIAB, Pakka Anna near Faisalabad. Gypsum stones are also seen which are used as amendment for mitigating the high SAR and RSC (Residual sodium carbonates) levels of the water

of this system is its shorter life time, relative to the tile drainage system. The mole system is ideal for managing surface water and can also be used to reclaim both saline and saline-sodic soils.

5.1.5 Vertical Drainage

Removing groundwater through pumping is the most effective method of lowering a high water-table (Plate 3.4). To be able to pump groundwater, there needs to be a pocket of very coarse sand or gravel (an aquifer) below the soil surface, into which a slotted pipe (a 'well-point' or 'spear') can be installed. Water drains into the pipes through the slots cut into it. The water is then pumped to the surface. These systems are also sometimes referred to as 'bores'. Apart from salinity and water-table control, pumping groundwater can also provide extra water to supplement irrigation supplies (depending on the salinity of the groundwater). Pumping groundwater from shallow aquifers (< 25 m) is the most effective way to alleviate salinity effects near the surface. In Pakistan, many tube wells have been developed through SCARP (Salinity Control and Reclamation Project) to lower the water-table in waterlogged or shallow (high) water-table areas, and they have successfully helped manage soil salinization.

6 Salinity Control and Methods of Irrigation

In arid and semi-arid zones salinization is common due to an annual rainfall which is insufficient to leach salts. Because of this, there are limited quantities of good quality water and this necessitates the use of saline water in agriculture.

In order to address the soil salinization in irrigated agriculture fields, it is important to select a suitable irrigation system based on the soil conditions, water salinity level, crop type and available resources. The correct choice can allow a farmer to manage irrigation-induced soil salinization at an acceptable level, without invoking salinity hazards to the soil. The reader is referred to Chap. 4 of this book to learn more about the available irrigation systems and salt accumulation in the soil. Irrigation systems are, thus, only briefly described below.

6.1 Surface Irrigation

Application of water by gravity flow to the soil surface is termed as surface irrigation, which includes flood, basin, border, and furrow methods. Irrigation applied by these methods develops salinity zones in soil based on the frequency and amount of water applied in each irrigation cycle. At the end of each irrigation cycle, the soil dries out and salts are concentrated. This adversely affects plant growth. Increasing the frequency of irrigation can lower the salinity but it may also waste water. Alternative methods to improve the efficiency of water include the drip or sprinkler irrigation systems, whereas nuclear technique such as using neutron moisture probe (Chap. 6) offers the best solution to use water efficiently under saline conditions. The change from surface irrigation to more modern irrigation systems is costly and will require justification, as well as better crop adaptability. Under a surface irrigation system, leaching is usually used to keep the salinity controlled in the root-zone.

6.2 Basin Irrigation

In basin irrigation, bunds are created around the field to prevent the water flowing out, thus, confining the irrigation water to the target area. This method is commonly practiced for rice cultivation (rice grown by ponding) and for trees. In the United Arab Emirates and other countries, date palms are grown in small basins, with the tree being planted in the center of the basin. It should be kept in mind that the basin method is most suitable for sandy soils where water leaches down fairly quickly. However, if the crops or trees are sensitive to ponding water, this method should be avoided. In basin irrigation system, surface salinity is controlled, although at the subsurface wetting zone soil salinity will develop.

6.3 Furrow Irrigation

In the furrow irrigation method, small channels are created in the field to carry water to the plants. When the water enters to the furrows some water infiltrates into the soil, the amount being based on the soil texture, and this water also flows along the slope. Under such an irrigation system, the crop plant is grown on the furrow ridges. Development of a salinity zone in the furrow system depends upon the furrow to be irrigated. If all furrows are used for irrigation, the maximum salinity development is on the center-top of the furrow ridge. If alternate furrows are used, then the salinity development zone is on the opposite side of the ridge. These potential salinity zones should be avoided when planting the seeds.

6.4 Border Irrigation

When border irrigation is to be used, the land is divided into different parcels of land each of which is surrounded by bunds to confine the water. The water is applied to the soil through small water channels. This practice is very common in India and Pakistan, where governments have taken the initiative to line the water channels to prevent water from seepage into the soil. Root-zone salinity can be controlled by using excess irrigation water to leach salts into a soil zone below the roots. However, if the soil is fine textured, capillary rise after an irrigation event can develop a salt crust at the soil surface.

6.5 Sprinkler Irrigation

A sprinkler irrigation system is similar to rainfall, i.e. water is sprinkled on the soil surface. The sprinkler system requires pipes to be buried in the soil at specific depth and water enters to these pipes for irrigation. Sprinkler systems often allow efficient and economic use of water and reduce deep percolation losses. Chhabra (1996) is of the view that, if water application through sprinkler is in close agreement with crop needs (evapotranspiration and leaching), drainage and high water table problems can be greatly reduced, which in turn should improve salinity control.

6.6 Drip Irrigation

The drip irrigation method is the most efficient among all modern irrigation methods. In this system, water is applied precisely to the plants on a daily basis to meet the water requirement of crops. Drip irrigation is also ideal for delivering nutrients to the root-zone, thus optimize nutrient use efficiency. Drip irrigation has a priority over sprinkler irrigation, as the latter may cause leaf burn, defoliation of sensitive species, which generally does not occur with drip irrigation. The system consists of plastic pipes with emitters at specific intervals which are based on the distance between the plants in the rows.

7 Biological Methods of Soil Reclamation

Biological methods of soil reclamation include the use of organic material(s) to improve soil structure and to mobilize calcium from calcium carbonates through the decomposition process. ***Biosaline Agriculture*** (growing salt tolerant crops) is also part of biological reclamation.

7.1 *Use of Organic Amendments*

The soils of the arid and semi-arid regions are generally deficient in organic matter where saline and sodic soils are commonly found. The dispersed sodium in soil degrades the soil structure and restricts root growth and water movement in soil. Under such conditions, it is essential to improve soil structure. The organic matter can be added in following ways.

- Mixing previous crop stubbles into the soil
- Addition of farm yard manure
- Addition of crop residues into the soil
- Use of mulch material(s)
- Growing of green manure crops, such as legumes

The addition of crop residues and other organic materials improves soil structure. The legume crops used for green manure, in addition to adding organic matter, also add nitrogen into the soil, thus providing a dual benefit. Development of soil structure prevents soil erosion and hastens soil reclamation, primarily due to increased infiltration. The decomposition of organic matter produces a high level of CO_2 and also increases organic acids (humic, fulvic), which lower the soil pH. These processes increase the solubility of calcium carbonate and mobilize calcium, thereby replacing exchangeable sodium from the soil exchange complex and reducing soil sodicity.

Organic amendments, when applied in conjunction with inorganic amendments, can be more effective (Dorgon et al. 1976). Awan et al. (2015) reported that application of farm yard manure alone or in combination with inorganic N fertilizer has significant effect on wheat yield on saline-sodic soil both under monoculture and in agro-forestry systems. The selection of the organic matter to be applied is very important in order to avoid causing N deficiency (adding an amendment which has too high C:N ratio) and also increase salinity (e.g., cow dung slurry). Use of green manure crops may have a better chance than farm yard manure of being successfully integrated into a soil reclamation management package.

7.2 *Biosaline Agriculture*

Very few plant species grow well on saline soils, most either fail to grow or their growth is appreciably retarded. Salinization, thus, restricts options in choosing a successful crop species. Biosaline agriculture is the economic utilization of salt-affected soils for agricultural purposes, e.g. the growing of salt tolerant crops of agricultural significance. Biosaline agriculture includes the use of salty water for sustained agriculture. In the past, the term saline agriculture was used; however, a broader definition is now needed, one which includes the manipulation of desert and sea resources for both food and fuel (energy) production. For a successful adoption of biosaline agriculture, the following two points should be considered.

- Locations where biosaline agriculture is to be practiced must be studied carefully and potential problems diagnosed
- Based on the diagnostics results, one must choose appropriate measures for maximizing economic returns under each specific situation

Biosaline agriculture has a wide scope with diversified dimensions. These include:

- Breeding for salt tolerance within appropriate plant species
- Selection of salt tolerant genotypes, i.e. 'cultivars'
- Domestication of salt tolerant plants for economically sound (but sustainable) exploitation of salt-affected lands
- Climate smart agricultural practices (land preparation, planting, irrigation and fertilization, etc.).

Plant physiological studies will identify physiological factors controlling yield under the marginally saline conditions, and utilize physiological differences between salt tolerant and salt sensitive genotypes with a view to developing selection criteria for salt tolerance.

7.3 *Screening Methods*

The screening of a range of crop varieties (cultivars) across different levels of salinities can be a useful way to begin moving toward biosaline agriculture. There are a number of screening methods and the best will closely simulate the conditions under which a crop variety will be grown. Techniques range from laboratory investigations of seed germination capabilities to glasshouse studies and field experiments, and are discussed in detail by Shahid (2002), and briefly below.

7.3.1 Screening in Greenhouse Using Hydroponics

The seeds of different crop cultivars are first germinated in small dishes, and at the 2–4 leaf stage, the plants are transferred to aerated Hoagland's solution by carefully placing them through small holes made in thermopol sheets (which float on the surface of the hydroponic culture solution). The salinity of culture solution is then increased stepwise, being maintained at a range of levels, e.g. 0, 100, 150, 300 mM NaCl, etc. The cultivars which survive at higher salt concentration undergo a preliminary selection, and are then subjected to further testing in both greenhouse and under field conditions.

7.3.2 Screening in the Field

Two field screening methods are commonly used.

In the first method, different varieties/cultivars of a crop are grown in lines. On one corner of the field, the sprinkler system is established with a non-saline, fresh water supply. In the other corner of the field, a sprinkler system is installed with a saline water supply. The water from both systems is sprinkled in different ratios through the use of special adjustments of the nozzles for different lines of the various cultivars. Plastic cups can be placed in each line of plants in order to collect samples of the 'mixed' water sprays for assessment of the salinity of water applied in the field. The shoot or whole plant dry matter and the grain yields are measured as a criterion of salt tolerance (Shahid 2002).

In the second method, different varieties/cultivars are grown in different lines. Each set of crop plants is irrigated through drip or sprinkler irrigation system with water of different salinities. As in the first method, grain and biomass yield can be used as a measure of salt tolerance of the differing crop varieties.

8 Serial Biological Concentration (SBC) Concept

The Serial Biological Concentration of Salts (SBCS) concept was introduced by Heuperman (1995). The SBC was developed as a multiple production system to utilize drainage water from irrigation schemes. In SBC, the drainage water of increasing salinity is collected and reapplied to 3 or more successive irrigation plots on which crops of known (different) salt tolerance are planted (Blackwell 2000; Cervinka et al. 1999). The SBC system involves the reuse of drainage water on progressively more salt tolerant crops. Each crop is underlain by a tile drain for the collection of water to be used to irrigate the next stage. Within the crop sequence, the drainage water collected is reduced in volume due to plant water use. Thus, the salinity of the drainage water increases since there will be little or no salt uptake by the plants. The final effluent water is contained in relatively small evaporation ponds. This makes it feasible to consider the use of a floor lining for the pond in order to eliminate leakage. These 'salt water' ponds could also be used for fish farming. The highly saline water can also be collected in a series of ponds where, through evaporation, the salts can be collected if they have commercial value, or need to be safely disposed off.

9 Genetic Engineering (Developing Salt Tolerant Cultivars)

Molecular biology and the use of appropriate methods of genetic engineering could also play a role in developing salt tolerant crop genotypes (varieties) which are resistant to marginal environments (drylands, saline lands) for food and/or biomass production. Shahid and Alshankiti (2013) has identified some researchable ideas (or areas) which could lead to solutions for meeting the food demand of the earth's growing human population. Researchable ideas that may help meet the need for sustainable increases in crop production (Shahid and Alshankiti 2013) are listed below.

- Develop cultivars which have low water requirement, and ones with stomatal closure midday (to reduce transpiration)
- Introduce a Biological Nitrogen Fixation (BNF) character in non-leguminous crops to reduce dependence on commercial N fertilizer
- Enhance sunlight use efficiency for photosynthesis, thereby yielding increased dry matter production
- Introduce resistance to heat shock, salinity and water stress, thereby yielding more drought tolerant varieties, and
- Develop viable options to maximize yield under warmer (and water deficit) conditions through traditional breeding and agronomic research

10 Crop Yield Estimation Under Saline Conditions

Crop yields decrease as a factor of increasing soil salinity above a threshold salinity value. Crops can tolerate salinity up to a certain level (Maas 1990) without a measurable loss in yield, i.e. the 'threshold salinity'. As a general rule, the more salt tolerant is the crop, the higher is the threshold salinity level. Crop yields are reduced in a linear manner as salinity increases above this threshold salinity, as shown in the equation.

$$\mathrm{Yr} = 100 - \mathrm{S}\ (\mathrm{ECe} - \mathrm{t})$$

Where, Yr is crop yield relative to the same conditions without salinity, t is the threshold salinity, S is the % linear rate of yield loss with a 1 ECe (dS m^{-1}) increase above the threshold value. ECe is the electrical conductivity of the soil saturation extract and represents the average root-zone salinity. The expected yield (Yr) of a crop grown at a specific level of salinity (ECe) can, thus, be calculated. The reader is referred to Chap. 4 of this book for further details.

11 Integrated Soil Fertility Management (ISFM)

In parallel to the management of salinity and sodicity in agriculture fields, it is equally important to keep the soils healthy and productive through the maintenance of an optimal soil fertility status. The soils of the arid and semi-arid regions of the world are inherently low in soil fertility. There is an ongoing need to replenish the soil's nutrients through strategic use of chemical fertilizers and organic manure (s) that will ensure sustainable yields. The ISFM is an effective strategy for sustainable agriculture.

The replenishment of soil nutrient pools, on farm recycling of nutrients, reducing nutrient losses and improving the efficiency of inputs on saline and sodic soils is much more important than on good quality non-saline soils. The ISFM combines the use of both organic and inorganic sources to increase crop yield, rebuild depleted soils and protect a wide range of natural resources. Organic amendments can increase the efficiency of inorganic fertilizers through positive interactions on soil biological, chemical and physical properties. The ISFM optimizes the effectiveness of fertilizer and organic inputs in crop production and its implementation can rehabilitate degraded soils and restore their sustainable productivity. To be successful in nutrient replenishment for sustainable crop production, a new **4R** strategy needs to be used.

11.1 What Is a Four Right (4R) Strategy?

The four R (4R) nutrient strategies should be used to offset the plant's nutrient requirement, which involves:

- Right type of chemical fertilizers – e.g., ammonium versus nitrate based fertilizers
- Right rate of fertilizer – based on soil testing and target yield
- Right time of fertilizer application at the right growth stage – apply each fertilizer when the plants need specific nutrients
- Right location of fertilizer application – apply to the root-zone area where the nutrient can best be absorbed by plants

The fertilizer use efficiency of nitrogen fertilizers under field conditions is assessed using isotopic techniques of nitrogen-15 (see Chapt. 6 for detailed information).

12 Conservation Agriculture (CA)

Conservation agriculture (CA) is part of Climate Smart Agriculture (CSA). CA recognizes the importance of the upper 0–20 cm of the soil as the most active zone, and also the zone most vulnerable to erosion and land degradation. By protecting this critical soil zone, we ensure the continuity of good agriculture and a good environment.

Main principles of conservation agriculture, discussed in Dumanski et al. (2006), are listed hereunder.

- Maintaining a permanent soil cover and making certain that there is a minimal mechanical disturbance of the soil through the use of zero tillage systems. This will help ensure sufficient living and/or residual biomass to enhance soil and water conservation and control soil erosion.
- Promoting a healthy, living soil through crop rotations, cover crops, and the use of integrated pest management technologies
- Promoting the application of appropriate fertilizers, pesticides, herbicides, and fungicides in a strategic way to maintain a sustainable balance with crop requirements
- Promoting precision placement of inputs to reduce farm costs, optimize efficiency of operations, and prevent environmental damage
- Promoting legume fallows (including herbaceous and tree fallows where suitable), composting and the use of manures and other organic soil amendments
- Promoting agro-forestry for fiber, fruit and medicinal purposes.

13 Climate Smart Agriculture (CSA)

Climate smart agriculture (CSA) includes proven practical techniques and approaches that can help achieve food security, adaptation to and mitigation of the effects of a changing climate. Increasing soil organic matter content and moisture through low to zero tillage and the use of mulching make crop yields more resilient and combat soil degradation. The introduction of integrated soil fertility management can also reduce chemical fertilizer costs.

CSA seeks to increase productivity in an environmentally and socially acceptable way, strengthen farmers' resilience to climate change, and reduce agriculture's contribution to climate change by reducing greenhouse gas emissions and increasing carbon sequestration and storage on farm land. Climate smart agriculture includes

Plate 3.5 Neglected but precious resource for salts in UAE (left) and Bahrain (right) needs attention

proven practical techniques such as mulching, intercropping, conservation agriculture, crop rotation, integrated crop-livestock management, agro-forestry, controlled grazing, and improved water management. It requires innovative practices such as better weather forecasting, early warning systems and risk insurance. It is also very much about getting existing technologies off the shelf and into the hands of farmers, as well as developing new technologies such as drought or flood tolerant crops to meet the demands of the changing climate. Finally, climate smart agriculture is about creating and enabling the policy and environment which will allow for adaptation (World Bank 2011).

14 Commercial Exploitation of Mineral Resources from Highly Saline Areas – The Neglected Resource

The lands of immediate vicinity to the coast are highly vulnerable to sea water intrusion. These soils, overtime, are converted to ***sabkha*** (salt scald) – areas which are not conducive for agricultural activities. Such highly saline lands may, however, be exploited for other uses, such as commercial salt harvesting. This precious resource (Plate 3.5) is neglected for the time being, but has high potential to generate capital.

In Australia, while addressing dryland salinity issues, an approach entitled '***Options for the productive use of salinity – OPUS***' has been successfully used in a National Dryland Salinity Program (PPK E & I Pty Ltd. 2001). One of the options is the industrial use of harvested salts.

The OPUS approach has the following objectives.

- Collate and assess information on innovative options for the productive use of saline land and water, both within Australia and internationally
- Provide guidance and considerations for industry (or industrial) implementation

- Assess the economic and marketing barriers to investment in industries involved in salinity issues
- Suggest the skills, resources and institutional arrangements that would improve our capacity to utilize saline resources and identify areas requiring further research and development.

The OPUS assessed 13 industries, including sheep grazing on saltbush '*Atriplex*' pastures, saline forestry, aquaculture of fish, algal production and desalination.

In this section, however, our emphasis is mainly on salt harvesting and exploitation of the salt(s) for commercial purposes. Sea water is dominant in Na^+ and Cl^- ions relative to other ions (Ca^{2+}, Mg^{2+}, $SO_4{}^{2-}$, $CO_3{}^{2-}$, $HCO_3{}^-$, etc.). Thus, practically speaking, when we talk of salt, it means sodium chloride (NaCl), the mineral name being 'halite'. The PPK E & I Pty Ltd. (2000) described three types of salts which are harvested in different ways. There are three broad categories:

Rock salt – subsurface deposits within the earth, formed millions of years ago, an era when the oceans that covered the planet evaporated and receded, leaving behind salt deposits. Rock salt is mined in the mineral form.

Solar salt – saline sea water is pumped into condensing ponds and then to the saturating ponds. The evaporation leads to salt crystallization, which is harvested. In the United Arab Emirates, sea water intrusion into the coastal areas and subsequent evaporation has developed huge quantities of salts which have the potential for commercial harvesting.

Evaporated salt – here, wells are drilled into underground salt deposits and water is pumped into the wells to dissolve the salts. The resulting brine is pumped to the surface, evaporated and harvested.

There are many uses of salts in the industry. Chemical industry accounts for 55% of global salts consumption. Three main products are:

(i) Caustic soda – NaCl (halite) + H_2O (water) → NaOH (Caustic soda) + HCl (Hydrochloric acid)
(ii) Soda ash, and (iii) Chlorine

There are a number of commercial uses of the above products in the pulp, paper, organic and inorganic chemicals, glass, petroleum, plastics (PCV) and textiles industries (IMC Global 1999; Olsson Industries 2001; Dampier Salt Pty Ltd. 2001; Cheetham Salt Pty Ltd. 2001).

Salts are used globally in the food industry, i.e. preserving and preparing canned and bottled foods, in cheese production, in bakeries and cooking foods, etc.

In Europe or other countries where heavy snow falls are frequent, salt is used to de-ice the roads to facilitate transport. Such a use accounts for 30% of the total salt use in Europe (European Salt Producers' Association 2000).

Plate 3.6 (**a**) *Eucalyptus camaldulensis* stand on a saline-sodic soil at the Biosaline Research Station of NIAB (Pakka Anna), near Faisalabad, monitored for tree water use with HeatPulse Data Loggers. (Adapted from Mahmood et al. 2004), (**b**) Setup for monitoring tree water use with HeatPulse Technique

15 Salinity Control Strategy

A salinity control strategy should be to stop the spread of salt-affected soils, with a major objective to greatly reduce the future effects of salinity. This requires a commitment by governments. This strategy/objective can be achieved in a number of ways, including:

- By re-vegetating the shallow water-table areas with deep-rooted trees, and
- Lowering the water-table by pumping, and flushing the salts from soils

Perennial plants and forages, especially alfalfa, are useful for lowering water-table, because they have a longer growing season and take up more water from a greater depth in the soil than annual plants. *Eucalyptus* has been used to lower water-table (***biodrainage)*** as it transpires large volumes of water. Mahmood et al. (2001) reported on the water use of *Eucalyptus* and other salt tolerant tree species, determined by using HeatPulse Data Loggers, under variable field conditions in Punjab province, Pakistan (Plate 3.6).

Eucalyptus camaldulensis on an irrigated, non-saline site near Lahore showed an annual water use of 1393 mm (Table 3.4). Irrigated *Eucalyptus microtheca* at this site and un-irrigated *E. camaldulensis* dependent on saline groundwater on saline soil at Pakka Anna near Faisalabad also transpired over 1000 mm of water per year. *Acacia ampliceps* showed much lesser water use than *E. camaldulensis* in spite of a similar basal area growth at Pakka Anna, whereas lowest annual water use of 235 mm was shown by an under-stocked stand of *Prosopis juliflora* at this site. These results provide an example regarding the range of choice of suitable tree species for site-specific conditions with reference to water availability and objectives of re-vegetation projects.

Table 3.4 Calculated daily and annual water use by plantations on saline sites near Lahore and Pakka Anna near Faisalabad, Pakistan. (Adapted from Mahmood et al. 2001)

Plot details	Soil $EC_{1:1}$ (dS/m)	Days monitored	Mean daily water use (mm) ± S.E.	Annual water use (mm)
Lahore				
Eucalyptus camaldulensis	2.5–5.0	333	3.82 ± 0.07	1393
E. microtheca	2.5–5.0	322	2.87 ± 0.06	1084
Pakka Anna near Faisalabad				
E. camaldulensis (low salinity)	3.2–4.0	330	3.20 ± 0.07	1169
E. camaldulensis (high salinity)	6.2–8.5	285	2.99 ± 0.09	1090
Acacia ampliceps	5.0–5.2	317	1.71 ± 0.05	624
Prosopis juliflora	6.1–7.0	262	0.64 ± 0.01	235

The strategy should direct the farming community's efforts to areas where salinity is, or will be, a major problem. The main emphasis should be to provide further encouragement, assistance and technical support to research scientists in order to identify the areas where the most effort should be directed. These areas, once identified, should be considered as '***hot spots***' and most of the resources should be directed into rectifying and preventing future enlargement of these hot spots.

Working together to tackle the salinity problems is sensible, especially when the cause may not necessarily be confined to one property. Governments should take appropriate steps in improving long-term productivity and amenity value of saline areas. Grants and incentives should be made available to educate the farming community, to make it aware of the land degradation problem and the need for farmers to move quickly and properly in order to protect their livelihood.

Education of the farming community is vital in increasing the community's awareness and understanding of salinity, so that the above strategy is widely supported and acted upon. Advisory programs should be developed using extension workers so that farmers can plan and use salinity control practices on their farms.

Salinity mapping on a whole farm scale is the best practice for crop selection. Salinity exhibitions for community education should be arranged in government institutes, and demonstration days at the farmers' fields are also useful. Preparation of introductory brochures for salinity control and management at the farm level and their distribution to the farming community can enhance the understanding of how to best tackle salinity in a sustainable way. The awareness of the problem by local school teachers can give students a hands-on experience and help them discuss options with their students. After all, today's students will be tomorrow's land managers.

References

Awan AR, Siddiqui MT, Mahmood K, Khan RA, Maqsood M (2015) Interactive effect of integrated nitrogen management on wheat production in *Acacia nilotica*- and *Eucalyptus camaldulensis*-based ally cropping systems. Int J Agric Biol 17:1270–1127

Ayers RS, Westcot DW (1985) Water quality for agriculture. FAO irrigation and drainage paper 29 rev 1. Food and Agriculture Organization of the United Nations, Rome, Italy, 174 pp

Blackwell J (2000) From saline drainage to irrigated production. Research project information from CSIRO land and water, sheet no. 18. Communication Group, CSIRO Land and Water, Griffith, Australia, 4 pp www.clw.csiro.au/division/griffith

Cervinka V, Diener J, Erickson J, Finch C, Martin M, Menezes F, Peters D, Shelton J (1999) Integrated system for agricultural drainage management on irrigated farmland, final research report 4-FG-20–11920, five points: Bureau of Reclamation, US Department of the Interior, Sacramento, California, USA, 41 pp

Cheetham Salt Pty Ltd (2001) Salt from the sea. Cheetham Salt website: www.cheethamsalt.com.au

Chhabra R (1996) Irrigation and salinity control. In: Chhabra R (ed) Soil salinity and water quality. Oxford and IBH Publishing Co Pvt Ltd, New Delhi/Calcutta, pp 205–237

Dampier Salt Pty Ltd (2001) Salt and its uses. Dampier salt website: www.dampiersalt.com.au

Dargan KS, Gaul BL, Abrol IP, Bhumbla DR (1976) Effect of gypsum, farmyard manure and zinc on the yield of barseem, rice and maize grown in highly sodic soil. Ind J Agric Sci 46:535–541

Dumanski J, Peiretti R, Benetis J, McGarry D, Pieri C (2006) The paradigm of conservation tillage. In: Proceedings of world association of soil and water conservation P1:58–64

European Salt Producers' Association (2000) Salt in the EU. Website: www.eusalt.com

Heuperman AF (1995) Salt and water dynamics beneath a tree plantation growing on a shallow watertable. Report of the department of agriculture, energy and minerals victoria, Institute for Sustainable Irrigated Agriculture, Tatura Center, Australia 61 pp

IMC Global (1999) World crop nutrients and salt situation report. IMC Global website: www.imcglobal.com

Maas EV (1990) Crop salt tolerance. In: Tanji KK (ed) Agricultural salinity assessment and management manual. ASCE manuals and reports on engineering no 71, ASCE New York, USA, pp 262–304

Mahmood K, Morris J, Collopy J, Slavich P (2001) Groundwater uptake and sustainability of farm plantations on saline sites in Punjab province, Pakistan. Agric Water Manage 48:1–20

Mahmood K, Hussain F, Iqbal MM (2004) Eucalyptus camaldulensis – a suitable tree for water-logged and saline wastelands. Nuclear Institute for Agriculture and Biology (NIAB), Faisalabad, 6 pp

Mashali AM (1995) Network on integrated soil management for sustainable use of salt-affected soil. Proceedings of the international symposium on salt-affected lagoon ecosystems ISSALE-95, 18–25 Sept 1995, Valencia, Spain, pp 267–283

Olsson industries (2001). Olsson's salt information page. Website: www.olssons.com.au

PPK E & I Pty Limited (2001) Options for the productive use of salinity. National Dryland Salinity Program, Australia, 249 pp (+Appendices 37 pp)

Rhoades JD (1974) Drainage for salinity control. In: Van Schilfgaarde J (ed) Drainage for agriculture. Amer Soc Agron Monograph No 17:433–462

Rhoades JD, Merrill SD (1976) Assessing the suitability of water for irrigation: theoretical and empirical approaches. In: Prognosis of salinity and alkalinity. FAO Soils Bulletin 31 Food and Agriculture Organization of the United Nations, Rome, Italy, pp 69–110

Schoonover WR (1952) Examination of soils for alkali. University of California, Berkley, California, USA (Mimeographed), pp 104–105 (USSL Staff 1954 Handbook No 60 ?)

Shahid SA (2002) Recent technological advances in characterization and reclamation of salt-affected soils in arid zones. In: Al-Awadhi NM, Taha FK (eds) New technologies for soil reclamation and desert greenery. Amherst Scientific Publishers, Amherst, pp 307–329

Shahid SA, Alshankiti A (2013) Sustainable food production in marginal lands – case of GDLA member countries. Int Soil Water Conserv Res 1:24–38

Shahid SA, Muhammed S (1980) Comparison of methods for determining gypsum requirement of saline-sodic soils. Bull Irrig Drain Flood Control Res Counc Pak 19(1–2):57–62

Shahid SA, Rahman KR (2011) Soil salinity development, classification, assessment and management in irrigated agriculture. In: Passarakli M (ed) Handbook of plant and crop stress. CRC Press Taylor & Francis Group, Boca Raton, pp 23–29

Shahid SA, Aslam Z, Hashmi ZH, Mufti KA (2009) Baseline soil information and management of a salt-tolerant forage project site. Eur J Sci Res 27(1):16–28

Shahid SA, Abdefattah MA, Omar SAS, Harahsheh H, Othman Y, Mahmoudi H (2010) Mapping and monitoring of soil salinization – remote sensing, GIS, modeling, electromagnetic induction and conventional methods – case studies. In: Ahmad M, Al-Rawahy SA (eds) Proceedings of the international conference on soil salinization and groundwater salinization in arid regions, vol 1. Sultan Qaboos University, Muscat, pp 59–97

Shahid SA, Taha FK, Ismail S, Dakheel A, Abdelfattah MA (2011) Turning adversity into advantage for food security through improving soil quality and providing production systems for saline lands: ICBA perspectives and approach. In: Behnassi M, Shahid SA, D D'Silva J (eds) Sustainable agricultural development: recent approaches in resources management and environmentally-balanced production enhancement. Springer, Dordrecht, pp 43–67.

Sharma OP, Gupta RK (1986) Comparative performance of gypsum and pyrites in sodic vertisols. Ind J Agric Sci 56:423–429

USSL Staff (1954) Diagnosis and improvement of saline and alkali soils. USDA handbook no 60 Washington DC, USA, 160 pp

World Bank (2011) Opportunities and challenges for climate-smart agriculture in Africa. 8 pp http://climatechange.worldbank.org/sites/default/files/documents/CSA_Policy_Brief_web.pdf

Chapter 4
Irrigation Systems and Zones of Salinity Development

Mohammad Zaman, Shabbir A. Shahid, and Lee Heng

Abstract Selection of suitable irrigation systems (drip-surface and subsurface, sprinkler, bubbler, furrow etc.) for irrigated agriculture is one way of improving water use efficiency and to manage root zone salinity. These irrigation systems develop salinity zones differently which needs to be understood for various reasons, such as where to place the seed for good germination and where to apply leaching to maintain the root zone salinity below crop threshold salinity level. In this chapter emphasis have been made to describe various irrigation systems and zones of salinity development under each system. In surface irrigation system (flood, surge, sprinkler, bubbler) the maximum salinity is developed in deeper layers based on the wetting front and the lowest salinity is at the surface. Drip irrigation is often preferred to sprinkler irrigation for species with a high sensitivity to leaf necrosis. In surface drip irrigation salts concentrate along the perimeters of the expanding wetting soil zone, with the lowest salt concentrations occurring in the immediate vicinity of the water source, the highest at the soil surface, and in the very center of any two drippers, i.e. at the boundary of the volume of wetted soil. In the subsurface drip irrigation, the salts continuously buildup at the soil surface through an upward capillary movement from the buried irrigation lines during growing season, therefore the concept of leaching requirement (LR) does not work specially to leach the salts from surface above the buried drip lines. In furrow irrigation system maximum salts accumulate in ridges of soil between the furrows. The salt accumulation in furrow irrigation using different bed shapes (flat top bed, sloping beds) is shown in different figures giving guidelines to the farmers to place seeds in safe zone to accomplish high germination rate. Following the salinity development zones, various methods of salinity management are described. Relative crop salinity tolerance rating is described briefly. Prediction of crop yield in salinized farms compared to non-saline farms is also described using Maas and Hoffman equation.

Keywords Irrigation systems · Sprinkler · Drip · Surge · Salinity development zones · Salinity tolerance · Maas and Hoffman

M. Zaman et al., *Guideline for Salinity Assessment, Mitigation and Adaptation Using Nuclear and Related Techniques*, https://doi.org/10.1007/978-3-319-96190-3_4

1 Introduction

In arid and semi-arid regions, the major constraints to agriculture are water and arable land scarcity, harsh climatic conditions, and poor water use efficiency. This often necessitates the use of saline/brackish water to partially supplement the normal water requirements of crops. In order to minimize the effects of saline water on salinity in the root-zone soil, a suitable irrigation method must be selected, one which does not raise soil salinity hazards. The irrigation method chosen for a particular farm/field is determined by the depth of irrigation water applied, water losses by leaching and runoff, zones of salt accumulation, and the uniformity of applying the irrigation water.

Broadly, surface irrigation systems can be divided into two main classes: ***gravity flow surface irrigation*** (flood, border, surge, furrow, etc.), and '***pressurized flow irrigation***'. The practice of surface irrigation is predominant and covers nearly 95% of the world's irrigated areas. The sustainability of surface irrigation depends on the use of innovative methods, ones which are appropriate for different irrigation systems and result in a wide adoption by farmers. Sprinkler and trickle irrigation together represent the broad class '*pressurized*' irrigation methods. In trickle irrigation, the water is carried in a pipe system to the point of irrigation, where the water is finally made available to the root system for uptake by plants. Surface irrigation can lead to heavy losses through leaching while being conveyed to (and at the point of) the irrigation site.

Each irrigation system develops salinity at a specific soil zone and, thus, needs to be carefully monitored. Shahid (2013) has recently introduced zones of soil salinity development for a range of different irrigation systems. Commonly used irrigation methods and the probable zones of soil salinity development are discussed here. In this context, safe zones with a relatively low salinity are suggested where seeds can be placed for germination, or where seedlings can be transplanted.

The zone of salt accumulation depends on the method of irrigation and seed bed shape. The irrigation systems used include:

- Flood irrigation
- Basin irrigation
- Border irrigation
- Surge irrigation
- Furrow irrigation
- Drip irrigation
  - Surface drip irrigation
  - Subsurface drip irrigation

Soil salinity development, i.e. the location and quantity of salts in each irrigation system is variable. In the flood, basin, border and sprinkler irrigation systems, the net water movement is downward when there is no high water-table. Under such circumstances, surface accumulation of salts is unlikely. Rather, the salt accumulates

Plate 4.1 Basin irrigation of date palm

at deeper soil layers based on the final 'wetted zone'. Each irrigation cycle dissolves surface salinity and then concentrates those salts at the final wetting zone. Here, then, there is lower surface salinity and an increase in the subsurface salinity.

At the end of each irrigation (flood, basin and border) cycle, the soil dries out and the salts are concentrated, adversely affecting the crop yield. Frequent irrigation may lower the salinity, but it wastes water. Alternatives which improve the efficiency of water use are drip or sprinkler irrigation. In the bubbler type of (basin) irrigation, a small fountain of water is applied to flood small basins dug around the tree base, or on the soil surface adjacent to individual trees. In the GCC countries, this system is commonly used to irrigate date palm trees (Plate 4.1).

This shift from conventional surface irrigation to a more modern irrigation system is costly and requires assurance on a high degree of crop adaptability. However, there are advantages in using modern irrigation system(s), especially when saline/brackish water must be used under hot desert conditions like that prevail in the Middle East, and parts of Australia and South East Asia. Frequent (twice daily) irrigation maintains a soil moisture level that does not fluctuate appreciably between wet and dry extremes. This residual moisture which remains in the soil between irrigation cycles keeps salts in a dilute solution, making it possible to use saline water – a situation which is problematic when irrigation occurs every second or third day.

Plate 4.2 Sprinkler irrigation in a demonstration plot of salt tolerant grass in Abu Dhabi Emirate

2 Sprinkler Irrigation

With sprinkler irrigation, strong streams of water are sprayed through the air to spread on the soil surface (Plate 4.2). A good sprinkler irrigation (SI) must meet all of the requirements of the crop for water, including evapotranspiration (ET). Irrigation by sprinkler allows efficient and economic use of water and reduces losses through deep percolation of water through the soil. If water applied via SI is in close agreement with crop needs (ET plus leaching), excessive drainage and high water-table problems can be greatly reduced, thus improving salinity control. Sprinkler irrigation can be accomplished through the use of fixed sprinklers or by a continually moving system, such as center-pivot, linear moving laterals, and other forms of travelling sprinklers. Special care should be exercised in selecting nozzle size, operating pressure and sprinkler spacing when using SI on fine textured soil (which will have low intake rates) to ensure uniform water application at low rates.

While sprinkler irrigation will uniformly distribute water, high wind can distort the distribution of water applied, thus affecting water use efficiency. Windbreaks around the edges of the farm can help to reduce the negative effects of strong wind.

The saline water applied with sprinkler can also cause leaf burn (necrosis) through salt injury (Plate 4.3). Leaf necrosis from sprinkler irrigation can occur when sodium exceeds 70 ppm, or chloride exceeds 105 ppm in irrigation water.

Plate 4.3 Salinity diagnostics in a grass field where sprinkler irrigation with saline water has caused necrosis (leaf burn)

Table 4.1 Susceptibility of crops to foliar injury[a] from saline sprinkler water

Na^+ or Cl^- concentrations (meq l^{-1}) which can cause foliar injury			
< 5	5–10	10–20	> 20
Almond	Grape	Alfalfa	Cauliflower
Apricot	Pepper	Barley	Cotton
Citrus	Potato	Corn	Sugar beet
Plum	Tomato	Cucumber	Sunflower
		Safflower	
		Sesame	
		Sorghum	

Source data (Maas 1986)
[a]Foliar injury is influenced by cultural and environmental conditions
Data presented is for general guidelines for day-time sprinkling (cf. Minhas and Gupta 1992)

Thus, quality of water must closely match the leaf burn tolerance of the crop plants. The leaves of many plants readily absorb Na^+, Ca^{2+}, and Cl^- when water is applied through sprinkler system. The susceptibility of foliar injury differs among plant species; it is related to leaves' characteristics and rate of ion absorption rather than salinity tolerance (Maas 1986). However, sprinkler irrigation applied at night, or during periods of high humidity can reduce or eliminate the problem of leaf necrosis. Relative susceptibility of crops to foliar injury (Maas 1986) is shown in Table 4.1. Finally, the high costs of establishing and operating a sprinkler irrigation system limit its adoption by smallholder subsistence farmers.

Fig. 4.1 Salinity zone profiles occurring under a wide range of irrigation methods: sprinkler, flood, basin (bubbler) and border irrigation systems (Shahid 2013)

Under sprinkler irrigation, the salinity buildup occurs in the subsurface soil (Fig. 4.1). Thus, the SI system is highly effective in leaching salts from the surface and providing a soil environment which is conducive for seed germination and early stage of plant growth.

3 Drip Irrigation

Drip irrigation system can supply the required quantity of water to the crop on a daily or periodic basis. Drip irrigation delivers water near each plant through pipes (usually plastic) and a series of closely spaced emitters (drippers). This leads to high water use efficiency. The flow rate of each dripper can be controlled from 1 to 4 + liters per hour. The use of drippers for application of poor quality water may give better crop yields due to an ability to maintain high soil moisture levels and replenish the water lost by ET on a daily basis. Drip irrigation is often preferred to sprinkler irrigation for species with a high sensitivity to leaf necrosis. However, because the diameters of the dripper openings are quite small, the evaporation of saline water at

Plate 4.4 Wetting zone and salinity buildup in drip irrigation system: (**a**) Wetted soil, (**b**) Salt accumulation in the center of drip lines where wetting zones meet

the end of the dripper opening can lead to clogging, which reduces (or completely stops) the discharge of irrigation water from individual drippers. Thus, drippers must be inspected periodically to prevent this problem.

In drip irrigation, salt accumulation occurs via two processes. First, the soil becomes saturated with saline water and solutes are spread throughout the soil, saturating neighboring voids (Plate 4.4). In the second process, which occurs between consecutive irrigation cycles, both evaporation of water from the soil and the uptake of water and nutrients by plants are occurring. Solutes, thus, become redistributed in the soil with a final buildup of salts resulting from the interaction of these two processes throughout the crop season. During drip irrigation, salts will concentrate below the soil surface along the perimeters of the expanding wetting soil zone. Prolonged soil drying, or interspersing long intervals between irrigation cycles, can lead to increasingly saline soil-water movement back towards the plant, thereby increasing the likelihood of plant damage. This can be managed by ensuring that irrigation volumes are sufficient to allow the movement of new irrigation water to always be away from the drippers.

Salts concentrate through water evaporation from the soil and also by plant uptake. As discussed above, salt accumulation occurs on the boundaries of the wetted soil volume (Plate 4.4a), with the lowest salt concentration occurring in the immediate vicinity of the water source (Fig. 4.2). Salt concentrations will be the highest at the soil surface, and in the very center of any two drippers, i.e. at the boundary of the volume of wetted soil (Plate 4.4b).

Special care must be exercised to avoid the negative effects of salts to plants, especially during light rains that can push the salts from the center of drip lines towards plants and into the root-zone. Therefore, irrigation should be continued on schedule unless the rain is heavy (50 mm or more), which is very rare in arid and semi-arid regions especially in hot desert environments such as GCC countries. However, when such heavy rains do occur, they are usually sufficient to leach salts to deeper layers, leaving the root-zone salt free.

Fig. 4.2 A typical pattern of salt accumulation occurring from surface drip irrigation

In summary, irrigating daily is usually sufficient to continuously move the moisture down, into deeper soil zones, thereby keeping the salt levels under control.

3.1 Salinity Management When Using Drip Irrigation

In an attempt to reduce salinity effect in the root-zone, an experiment was conducted at ICBA experimental station to check the performance of drip irrigation (without a crop) at different dripper (drip emitter) spacings (25, 50 and 75 cm) using a saline water of 30 deci Siemens per meter (dS m^{-1}).

Soil samples collected from the centers of the emitters, were analyzed for electrical conductivity of soil extract from saturated paste (ECe). The ECe was recorded as 26 dS m^{-1} (25 cm spacing), 90 dS m^{-1} (50 cm spacing) and 102 dS m $^{-1}$ (75 cm spacing). The effects of emitters' spacing on soil salinity contours (top view) can be seen at a glance (Fig. 4.3). The larger the white areas became, the higher the soil salinity.

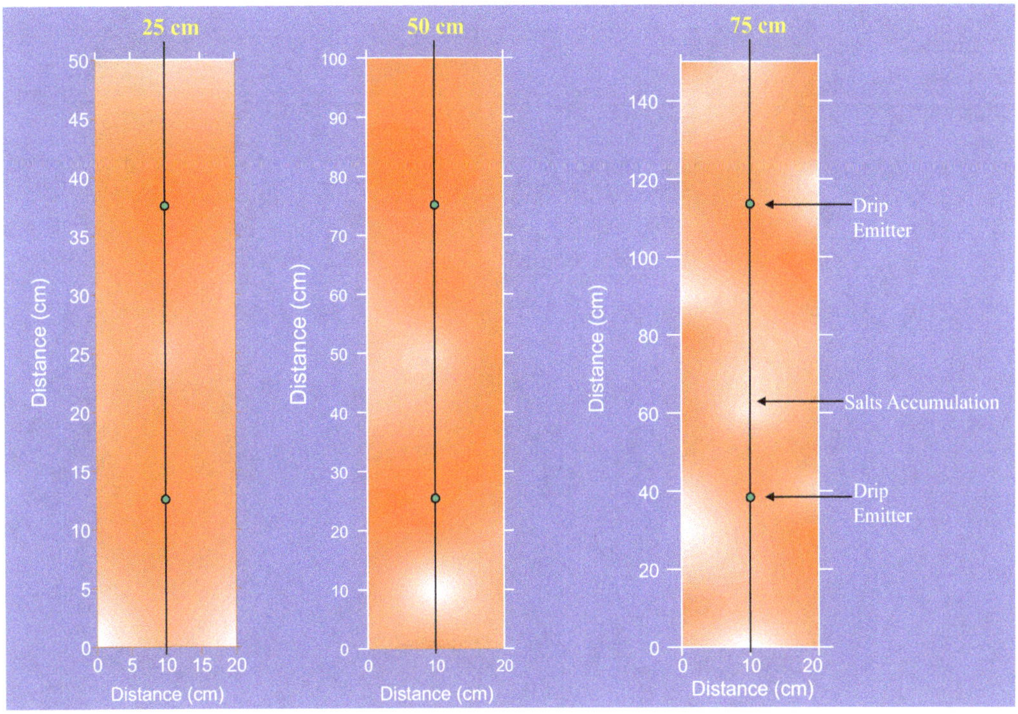

Fig. 4.3 Soil salinity under drip irrigation with emitter spacing at 25, 50 and 75 cm. Intensity of whiteness indicates higher salinity (Shahid and Hasbini 2007)

3.2 Subsurface Drip Irrigation

Subsurface drip irrigation (SDI) system, when compared with other irrigation systems, reduces water losses due to evaporation and deep percolation, while completely eliminating surface runoff (Phene 1990). The subsurface drip irrigation also increases marketable crop yield and quality (Ayers et al. 1999), while resulting in high nutrient use efficiency as well (Thompson et al. 2002).

The major limitation of SDI is the fact that salts continuously buildup at the soil surface through an upward capillary movement (Fig. 4.4) from the buried irrigation lines during growing season (Oron et al. 1999). This occurs because there is no above-soil water source, i.e. there is no way for irrigation water to leach the salts. The concept of leaching requirement (LR) does not function under subsurface drip irrigation specially to leach the salts from surface above the buried drip lines. However, salt accumulation in this zone above the buried irrigation line can be managed by supplementing subsurface drip irrigation with sprinkler irrigation (Thompson 2010). This approach may be costly, but is a necessary compromise. Salt accumulation occurs more rapidly when saline/brackish water is used, and also when the soils are fine textured. Only a heavy rainfall and/or occasional switch over from subsurface drip irrigation to sprinkler irrigation can leach salts from this zone. The alternative will be an accumulation of salts to toxic levels.

Fig. 4.4 Relative salt accumulation in the soil from subsurface drip irrigation showing high surface salinity in the zone above the irrigation line (Shahid 2013)

4 Furrow Irrigation

Furrow irrigation is most commonly practiced where soils are fine textured. In water scarce regions, and where the soils are sandy (such as GCC countries), furrow irrigation is not recommended. For farmers who do select furrow irrigation, there are various bed shape options to reduce salinity effects on plants (Bernstein et al. 1955; Bernstein and Fireman 1957; Bernstein and Francois 1973; Chhabra 1996) as described in the following sections.

In furrow irrigation, soil salinity varies widely from the base of the furrows to the tops of the ridges. Plate 4.5 shows salt accumulation in ridges of soil between the furrows. This pattern guides the best seed (or seedling) placement to minimize salinity effects, thereby achieving a higher crop yield. Re-plowing the furrow field for each new crop will redistribute the accumulated salinity, thereby allowing a continued cultivation in the area.

If a flatbed is chosen and both (two) furrows are irrigated, the zone of maximum salt accumulation will be in the center of the bed (Plate 4.5, Figs. 4.5, 4.6). In this case, it is safe to place the seeds or transplant seedlings away from the salt accumulation zone (Plate 4.5b). If, however, the farmer has chosen to place the seeds or transplant seedlings in a zone of maximum salt accumulation, it is highly

Plate 4.5 Pattern of salt accumulation (**a**), and safe zone for seed placement or transplanting (**b**) in a furrow irrigation system

Fig. 4.5 Salt accumulation when both furrows are irrigated; any plants growing in the very high salt accumulation zone will be affected

likely that either the seeds will not germinate or the seedlings will die over time (Fig. 4.7).

If alternate furrows are irrigated, the maximum zone of salt accumulation will be on the sides of the un-irrigated furrow (Fig. 4.8). In this situation, it is safe to place the seed or transplant seedlings away from the salt accumulation zone.

Fig. 4.6 Furrow irrigation system (flatbed); both furrows are irrigated

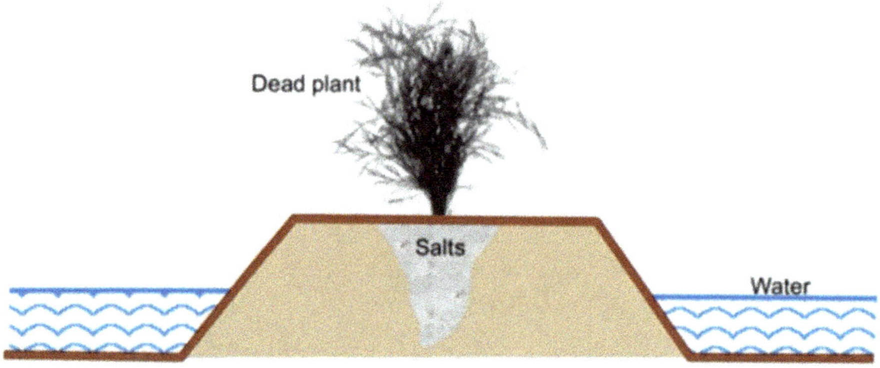

Fig. 4.7 Planting in the salt accumulation zone will result in a dead plant

Fig. 4.8 Salt accumulation and the safe zone for seeding when only the alternate furrow is irrigated

If a sloping bed is chosen, and depending upon the bed shape, the maximum salt accumulation will be either on the sides (Fig. 4.9) or in the center of the bed (Fig. 4.10). Avoid this zone of high salt accumulation, and place the seed or transplant the seedlings in the safe zone.

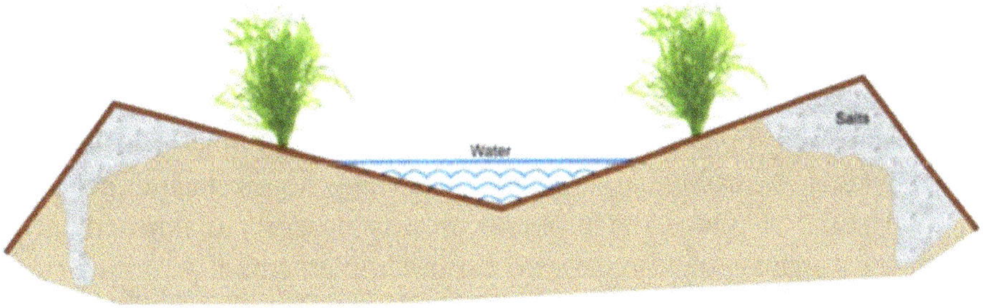

Fig. 4.9 Salt accumulation on sloping beds and the safe zone for seeding

Fig. 4.10 Salt accumulation on sloping beds. Note the safe zone for seeding when both furrows are irrigated

5 Surge Irrigation

Surge irrigation is a method of reducing the amount of runoff and allowing for a more uniform infiltration of irrigation water (Yonts and Eisenhauer 2008). It has long been recognized that water moves to the end of an irrigated field more quickly when applied intermittently than when applied continuously. In the latter case, and especially for coarse textured soils, it is practically impossible for continuously applied irrigation water to reach to the other end of the field; most of the water infiltrates into the soil at the water entrance end of the field.

How Surge Irrigation Works? When water first contacts the soil in the furrow, the infiltration rate is high; as the water flow continues, the infiltration rate is reduced to a near-constant rate. If water is shut off and allowed to infiltrate, a surface seal develops and when water is reintroduced, the infiltration rate into the previously wetted soil is reduced due to this partial sealing action. The end result is more water movement down the furrow and less infiltration into the soil. However, where soils are predominantly sandy, the surge irrigation method may not be a good option. In the GCC countries, surge irrigation has not gained recognition due to irrigation water scarcity, sandy soils and very hot climatic conditions.

6 Salinity and Sodicity Management in the Root-Zone

There is no single or universal technique to manage root-zone salinity. However, scientific diagnostics approach (Plate 4.6) based on a combination of engineering, chemical, physical, hydrological, biological and agronomic techniques can often yield a good solution. Once the problem area is properly diagnosed, a suitable selection of 'best management practices' can be implemented. A summary of such an approach is given below.

6.1 Physical Methods

Laser Guided Land Leveling – an improvement in leveling (preferably laser guided leveling) allows for a more uniform distribution of water.

Subsoiling – is 'deep ripping' to improve soil properties at deeper layers where a dense soil layer (or hard pan) exists, thereby limiting the penetration of roots and water infiltration.

Salts Scraping – salts at the soil surface can be scraped and removed to avoid further effects on plants after rain.

Plate 4.6 Soil sampling for root-zone soil salinity diagnostics

Sanding – sand can be added to a very fine textured (clayey) soil to improve soil texture, however, this practice can be very expensive and is impractical on a large scale basis.

6.2 Chemical Methods

Use of Amendments – It should be noted that salinity cannot be managed by chemical methods, but sodicity can be, and its management may have indirect effects on soil salinity. The most commonly used amendment to rectify soil sodicity is '*gypsum* ($CaSO_4.2H_2O$)', and the amount of gypsum to be applied will be based on the 'gypsum requirement' determined by standard laboratory methods. If, however, the soil contains sufficient quantities of $CaCO_3$ equivalents, then other amendments such as Sulfur (S), sulfuric acid (H_2SO_4) or pyrite (FeS_2), etc. can be used, again based on the 'gypsum requirement'. These amendments mobilize calcium from calcium carbonates equivalents and, thus, behave like gypsum to reclaim soil sodicity.

6.3 Hydrological Methods

Drainage System – a drainage system (surface and subsurface) can lower the soil water-table to a safer level in order to avoid detrimental effects of excess water in the normal plant root-zone. At the farm level, drainage is a 'moisture control system' that is required to maintain moisture and regulate salt balance in the root-zone.

Irrigation System – an irrigation system, when adopted, should permit frequent, uniform and efficient water application with as minimum a percolation loss as possible, but without curtailing essential leaching requirement. In addition, a good irrigation system should also avoid using saline water at the seed germination stage (a very sensitive stage). Where appropriate, and good quality water is also available, farmers should practice the use of re-cycled water for irrigation.

Leaching Requirement – where necessary, farmer should use water additional to the volume required for crop ET (evapotranspiration). This will allow salts to be leached down, below the root-zone.

The uses of saline/brackish water usually raise root-zone soil salinity. This salt accumulation can be controlled by applying water additional to the ET water requirement of the crop. This extra water will usually push the salts below the root-zone. The amounts of water required for leaching (leaching requirement – LR) can be calculated by standard procedures (Ayers and Westcot 1985).

$$LR = \frac{EC_w}{(5EC_e - EC_w)}$$

Where,

LR = leaching requirement ratio
ECw = EC of the irrigation water (dS m^{-1})
ECe = estimated EC of the average saturation extract of the soil root-zone profile for an appropriate yield (10%) reduction (dS m^{-1}) as presented by Ayers and Westcott (1985)

Example
Calculate leaching requirement for a sprinkler irrigation (SI) system for an alfalfa crop when irrigation water salinity is 5 dS m^{-1}.

The ECe that would give a 10% crop yield reduction is 3.4 dS m^{-1}, assuming threshold salinity level of alfalfa is 2 dS/m (ECe).

Using the above equation,

$$LR = \frac{5}{[(5 \text{ x } 3.4) - 5]} = 0.41$$

6.4 *Agronomic Methods*

Proper Seeding – use planting procedures that minimize the effects of salts on the seeds at germination and early plant growth stages (see earlier section on irrigation systems and salinity zones).

6.5 *Biological Methods*

Where it is not possible to practice conventional agriculture due to unavailability of good quality water, harsh environmental conditions and exceptionally saline lands, as a compromise the use of salt tolerant crops (***Biosaline Agriculture***) can be adopted. Table 4.2 provides guidelines for selection of crops tolerant to salinity.

7 Relative Crop Salinity Tolerance Rating

Relative crop salinity tolerance rating based on Fig. 4.11 is divided into five categories. Each group represents the crops with similar tolerance. Based on the data in Table 4.2, minimum and maximum ECe boundaries can be assigned to each

Table 4.2 Relative crop salinity tolerance rating

Relative crop salinity tolerance rating	Soil salinity (ECe, dS m^{-1}) at which yield loss begins
Sensitive (S)	< 1.3
Moderately sensitive (MS)	1.3–3.0
Moderately tolerant (MT)	3.0–6.0
Tolerant (T)	6.0–10.0
Unsuitable for most crops (unless reduced yield is acceptable)	> 10.0

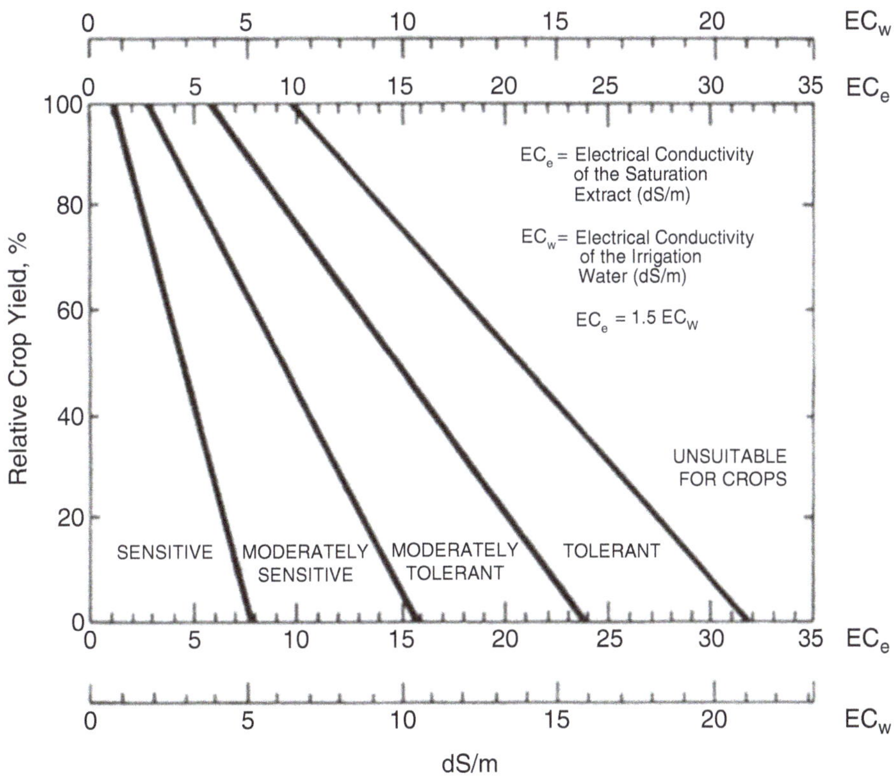

Fig. 4.11 Divisions for relative salt tolerance ratings of agricultural crops (Maas 1987). (Source: Ayers and Westcot 1985)

category. It should be noted that this broader division is for general guidelines and not meant to be a strict rule (Maas 1987).

8 Soil Salinity and Relative Yield Reduction of Crops

Crops can tolerate salinity up to certain levels without a measurable loss in yield (this is called the threshold level). As a general rule, the more salt tolerant is the crop, the higher is the threshold level. At salinity levels greater than the threshold, crop yield is reduced in a linear fashion as salinity increases. Using the salinity values from a salinity/yield model developed by Maas and Hoffman (1977), predictions of expected yield loss can be made (Table 4.3), as expressed in the following relationship.

$$Y_r = 100 - s\,(EC_e - t)$$

Where,

Y_r = percentage yield of the crop grown in saline conditions, relative to that yield obtained under non-saline conditions
t = threshold salinity level where the yield decrease begins
s = percent yield loss per increase of 1 ECe (dS m^{-1}) in excess of t

Salinity mapping at the farm level and the use of Table 4.3 may be used as a guide to predict yield losses.

General groupings for salt tolerance are shown in the schematic diagram in Fig. 4.11. The relative tolerance ratings, even if based on a limited amount of data, can be useful for comparisons among crops.

Table 4.3 Salt tolerance of important crops (Ayers and Westcot 1985)

Crop common name	Botanical name	Threshold (t) ECe, dSm^{-1}	Slope (s) % per dSm^{-1}	Rating[a]	Minimum[b] ECe, dSm^{-1}	Maximum[c] ECe, dSm^{-1}
Field crops						
Barley (forage)	*Hordeum vulgare*	8.0	5.0	T	8.0	28.0
Sugar beet	*Beta vulgaris*	7.0	5.9	T	7.0	24.0
Sorghum	*Sorghum bicolor*	6.8	16.0	MT	6.8	13
Triticale	*X Triticosecale*	6.1	2.5	T	6.1	46.0
Wheat	*Triticum aestivum*	6.0	7.1	MT	6.0	20.0
Wheat, durum	*Triticum turgidum*	5.9	3.8	T	5.7	20.0
Alfalfa	*Medicago sativa*	2.0	7.3	MS	2.0	16.0
Corn (maize)	*Zea mays*	1.7	12.0	MS	1.7	10.0
Cow peas	*Vigna unguiculata*	4.9	12.0	MT	4.9	13.0

(continued)

Table 4.3 (continued)

Crop common name	Botanical name	Threshold (t) ECe, dSm^{-1}	Slope (s) % per dSm^{-1}	Rating[a]	Minimum[b] ECe, dSm^{-1}	Maximum[c] ECe, dSm^{-1}
Vegetables						
Broccoli	*Brassica oleracea botrytis*	2.8	9.2	MS	2.8	14.0
Tomato	*Lycopersicon esculentum*	2.5	9.9	MS	2.5	13.0
Cucumber	*Cucumis sativus*	2.5	13.0	MS	2.5	10.0
Spinach	*Spinacia oleracea*	2.0	7.6	MS	2.0	15.0
Celery	*Apium graveolens*	1.8	6.2	MS	1.8	18.0
Cabbage	*Brassica oleracea capitata*	1.8	9.7	MS	1.8	12.0
Potato	*Solanum tuberosum*	1.7	12.0	MS	1.7	10.0
Pepper	*Capsicum annuum*	1.5	14.0	MS	1.5	8.5
Lettuce	*Lactuca sativa*	1.3	13.0	MS	1.3	9.0
Radish	*Raphanus sativus*	1.2	13	MS	1.2	8.9
Onion	*Allium cepa*	1.2	16.0	S	1.2	7.4
Carrot	*Daucus carota*	1.0	14.0	S	1.0	8.1
Beans	*Phaseolus vulgaris*	1.0	19.0	S	1.0	6.3
Turnip	*Brassica rapa*	0.9	9.0	MS	0.9	12.0
Fruits						
Date palm	*Phoenix dactylifera*	4.0	3.6	T	4.0	32.0
Orange	*Citrus sinensis*	1.7	16.0	S	1.7	8.0
Peach	*Prunus persica*	1.7	21.0	S	1.7	6.5
Apricot	*Prunus armeniaca*	1.6	24.0	S	1.6	5.8
Grape	*Vitus sp.*	1.5	9.6	MS	1.5	12.0

(continued)

Table 4.3 (continued)

Crop common name	Botanical name	Threshold (t) ECe, dSm^{-1}	Slope (s) % per dSm^{-1}	Rating[a]	Minimum[b] ECe, dSm^{-1}	Maximum[c] ECe, dSm^{-1}
Almond	*Prunus dulcis*	1.5	19.0	S	1.5	6.8
Plum, prune	*Prunus domestica*	1.5	18.0	S	1.5	7.1
Blackberry	*Rubus sp.*	1.5	22.0	S	1.5	6.0
Strawberry	*Fragaria sp.*	1.0	33.0	S	1.0	4.0

Adapted from Ayers and Westcot (1985); Maas (1990); Maas and Hoffman (1977)
S sensitive, *MS* moderately sensitive, *T* tolerant, *MT* moderately tolerant
[a]Relative crop salinity tolerance rating (see Table 4.2)
[b]Minimum ECe does not reduce yield (threshold)
[c]Maximum ECe reduces yield to zero

References

Ayers RS, Westcot DW (1985) Water quality for agriculture. FAO irrigation and drainage paper 29 rev 1. Food and Agriculture Organization of the United Nations, Rome, Italy, 174 pp

Ayers JE, Phene CJ, Humacher RB, Davis KR, Schoneman RA, Vail SS, Mead RM (1999) Subsurface drip irrigation of row crops: a review of 15 years of research at the water management research laboratory. Agric Water Manag 42:1–27

Bernstein L, Fireman M (1957) Laboratory studies on salt distribution in furrow-irrigated soil with special reference to the pre-emergence period. Soil Sci 83:249–263

Bernstein L, Francois LE (1973) Comparison of drip, furrow and sprinkler irrigation. Soil Sci 115:73–86

Bernstein L, Fireman M, Reeve RC (1955) Control of salinity in the Imperial Valley. California. USDA, ARS-41-4, 16 pp

Chhabra R (1996) Irrigation and salinity control. In: Chhabra R (ed) Soil salinity and water quality. Oxford and IBH Publishing Co Pvt Ltd, New Delhi, pp 205–237

Maas EV (1986) Salt tolerance of plants. Appl Agric Res 1:12–26

Maas EV (1987) Salt tolerance of plants. In: Christie BR (ed) Handbook of plant science in agriculture. CRC Press, Boca Raton, pp 57–75

Maas EV (1990) Crop salt tolerance. In: Agricultural salinity assessment and management. American Society of Civil Engineers, New York

Maas EV, Hoffman GJ (1977) Crop salt tolerance – current assessment. J Irrig Drain Div, ASCE 103(IR2):115–134

Minhas PS, Gupta RK (1992) Quality of irrigation water: assessment and management. Indian Council of Agricultural Research, New Delhi, 123 pp

Oron G, Demalach Y, Gillerman L, David I, Rao VP (1999) Improved saline-water use under subsurface drip irrigation. Agric Water Manag 39(1):19–33

Phene CJ (1990) Drip irrigation saves water. Proceedings of the National Conference and exposition. Offering water supply solution for the 1990's. Phoenix, USA, pp 645–650

Shahid SA (2013) Irrigation-induced soil salinity under different irrigation systems – assessment and management, short technical note. Clim Chang Outlook Adapt: Int J 1(1):19–24

Shahid SA, Hasbini B (2007) Optimization of modern irrigation for biosaline agriculture. Arab Gulf J Sci Res 25(1/2):59–66

Thompson TL (2010) Salinity management with subsurface drip irrigation. Proceedings of the international conference on soils and groundwater salinization in arid countries. 11–14 January 2010, Sultan Qaboos University, Muscat, Oman, vol 1, pp 9–13

Thompson TL, Doerge TA, Godin RE (2002) Subsurface drip irrigation and fertigation of broccoli: II. Agronomic, economic and environmental outcomes. Soil Sci Soc Am J 66:178–185

Yonts CD, Eisenhauer DE (2008) Fundamentals of surge irrigation. NebGuide University of Nebraska Lincoln – Extension Institute of Agricultural and Natural Resources Index: Irrigation Operations and Management July 2008. http://extensionpublications.unl.edu/assets/html/g1870/build/g1870.htm

Chapter 5
Irrigation Water Quality

Mohammad Zaman, Shabbir A. Shahid, and Lee Heng

Abstract The quality of irrigation waters differs in various regions, countries and locations based on how the groundwater has been extracted and used, the rainfall intensity and subsequent aquifer recharge. The use of groundwater for agriculture in hot arid countries where rainfall is scarce leads to increase groundwater salinity and limits the selection of crops for cultivation. It is therefore important to determine the irrigation water quality. The concentration and composition of soluble salts in water determines its quality for irrigation. Four basic criteria for evaluating water quality for irrigation purposes are described, including water salinity (EC), sodium hazard (sodium adsorption ratio-SAR), residual sodium carbonates (RSC) and ion toxicity. Toxicities of boron and chlorides to plants are described. More specifically the relative tolerance levels of plants to boron is tabulated for easy understanding. The most important part of this chapter is the modification of water quality diagram of US Salinity Laboratory Staff published in the year 1954, this diagram does not present EC over 2250 μS cm^{-1}, however, most of the irrigation waters present salinity levels higher than 2250 μS cm^{-1}. Therefore, to accommodate higher water salinity levels the water classification diagram is extended to water salinity of 30,000 μS cm^{-1} allowing the users of the diagram to place EC values above 2250 μS cm^{-1}. The salinity and sodicity classes are included in this chapter to provide information for crop selection and develop salinity and sodicity management options. The procedures for water salinity reduction through blending of different waters and management of water sodicity using gypsum are described by giving examples.

Keywords Irrigation · Quality · Salinity · Sodicity · Boron · Chlorides · Toxicities · Blending · Gypsum requirement

1 Introduction

Water scarcity is seen as a major constraint to intensify agriculture in a sustainable manner as an attempt to meet the food requirements of a rapidly growing human population. The ever increasing human population, climate change due to increased emissions of greenhouse gases (GHGs), and intensification of agriculture, are putting

M. Zaman et al., *Guideline for Salinity Assessment, Mitigation and Adaptation Using Nuclear and Related Techniques*, https://doi.org/10.1007/978-3-319-96190-3_5

severe pressure on the world's two major non-renewable resources of soil and water, and thus pose a big challenge to produce sufficient food to meet the current food demand. The present world population of 7.3 billion people is predicted to grow to over 9 billion by 2050, with the majority of this population increase occurring in developing countries, most of which already face food shortages. A 70% increase in current agricultural productivity will be required to produce sufficient food if these human population growth predictions prove to be correct. In this context, concerted efforts are being made globally to improve the effectiveness of water which will be used for enhancing the production of irrigated crops. Additionally, efforts are also being made to improve water harvesting and water conservation in rain-fed agriculture.

The injudicious use of saline/brackish water is all too often associated with the development of soil salinity, sodicity, ion toxicity, and groundwater pollution. Because of these negative effects, it is important to have a better understanding of exactly how the quality of water influences the management of irrigated agriculture, especially in arid and semi-arid regions.

Salinity, sodicity and ion toxicity are major problems in irrigation waters. In arid areas, where rainfall does not adequately leach salts from the soil, an accumulation of salts will occur in the crop's root-zone. Thus, periodic testing of soils and waters is required to monitor any change in salt content. Sodicity, the presence of excess sodium, will result in a deterioration of the soil structure, thereby reducing water penetration into and through the soil. Toxicity refers to the critical concentration of some salts such as chloride, boron, sodium and some trace elements, above which plant growth is adversely affected by those salts.

This chapter addresses several aspects of irrigation water quality and criteria to determine water quality. It will also cover management issues and soil responses to the use of irrigation water of varying quality. The information presented in this chapter is an updated and improved version of an excerpt from an earlier irrigation water quality manual (Shahid 2004).

2 Quality of Irrigation Water

The concentration and composition of soluble salts in water will determine its quality for various purposes (human and livestock drinking, irrigation of crops, etc.). The quality of water is, thus, an important component with regard to sustainable use of water for irrigated agriculture, especially when salinity development is expected to be a problem in an irrigated agricultural area.

There are four basic criteria for evaluating water quality for irrigation purposes:

- Total content of soluble salts (salinity hazard)
- Relative proportion of sodium (Na^+) to calcium (Ca^{2+}) and magnesium (Mg^{2+}) ions – sodium adsorption ratio (sodium hazard)

- Residual sodium carbonates (RSC) – bicarbonate (HCO_3^-) and carbonate (CO_3^{2-}) anions concentration, as it relates to Ca^{2+} plus Mg^{2+} ions.
- Excessive concentrations of elements that cause an ionic imbalance in plants or plant toxicity.

In order to achieve the first three important criteria, the following characteristics need to be determined in the irrigation waters: electrical conductivity (EC), soluble anions (CO_3^{2-}, HCO_3^-, Cl^- and SO_4^{2-}) where Cl^- and SO_4^{2-} are optional and soluble cations (Na^+, K^+, Ca^{2+}, Mg^{2+}) where K is optional. Finally, boron level must also be measured. The pH of the irrigation water is not an acceptable criterion of water quality because the water pH tends to be buffered by the soil, and most crops can tolerate a wide pH range. A detailed description of the techniques commonly employed for the analysis of irrigation water is available (USSL Staff 1954; Bresler et al. 1982).

2.1 *Salinity Hazard*

Excess salt increases the osmotic pressure of the soil solution, a situation that can result in a physiological drought condition. Thus, even though the soil in the field appears to have plenty of moisture, the plants will wilt. This occurs because the plant roots are unable to take up soil-water due to its high osmotic potential. Thus, water lost from the plant shoot via transpiration cannot be replenished, and wilting occurs.

The total soluble salts (TSS) content of irrigation water is measured either by determining its electrical conductivity (EC), reported as micro Siemens per centimeter ($\mu S\ cm^{-1}$), or by determining the actual salt content in parts per million (ppm). Table 5.1 prescribes the guidelines for water use relative to its salt content.

Table 5.1 Salinity hazard of irrigation water (Follett and Soltanpour 2002; Bauder et al. 2011)

Hazard	Dissolved salt content	
	ppm	EC ($\mu S\ cm^{-1}$)
None – Water for which no detrimental effects will usually be noticed.	500	750
Some – Water that may have detrimental effects on sensitive crops.	500–1000	750–1500
Moderate – Water that may have adverse effects on many crops, thus requiring careful management practices.	1000–2000	1500–3000
Severe – Water that can be used for salt tolerant plants on permeable soils with careful management practices.	2000–5000	3000–7500

2.1.1 Modified USSL Staff (1954) Water Salinity Classification

The USSL Staff (1954) water classification diagram does not present an EC over 2250 μS cm^{-1}. However, most of the water used for irrigation purposes possesses salinity levels which are higher than 2250 μS cm^{-1}. Therefore, in order to accommodate higher water salinity levels, Shahid and Mahmoudi (2014) have modified the USSL Staff (1954) water classification diagram by extending water salinity up to 30,000 μS cm^{-1} (Fig. 5.1).

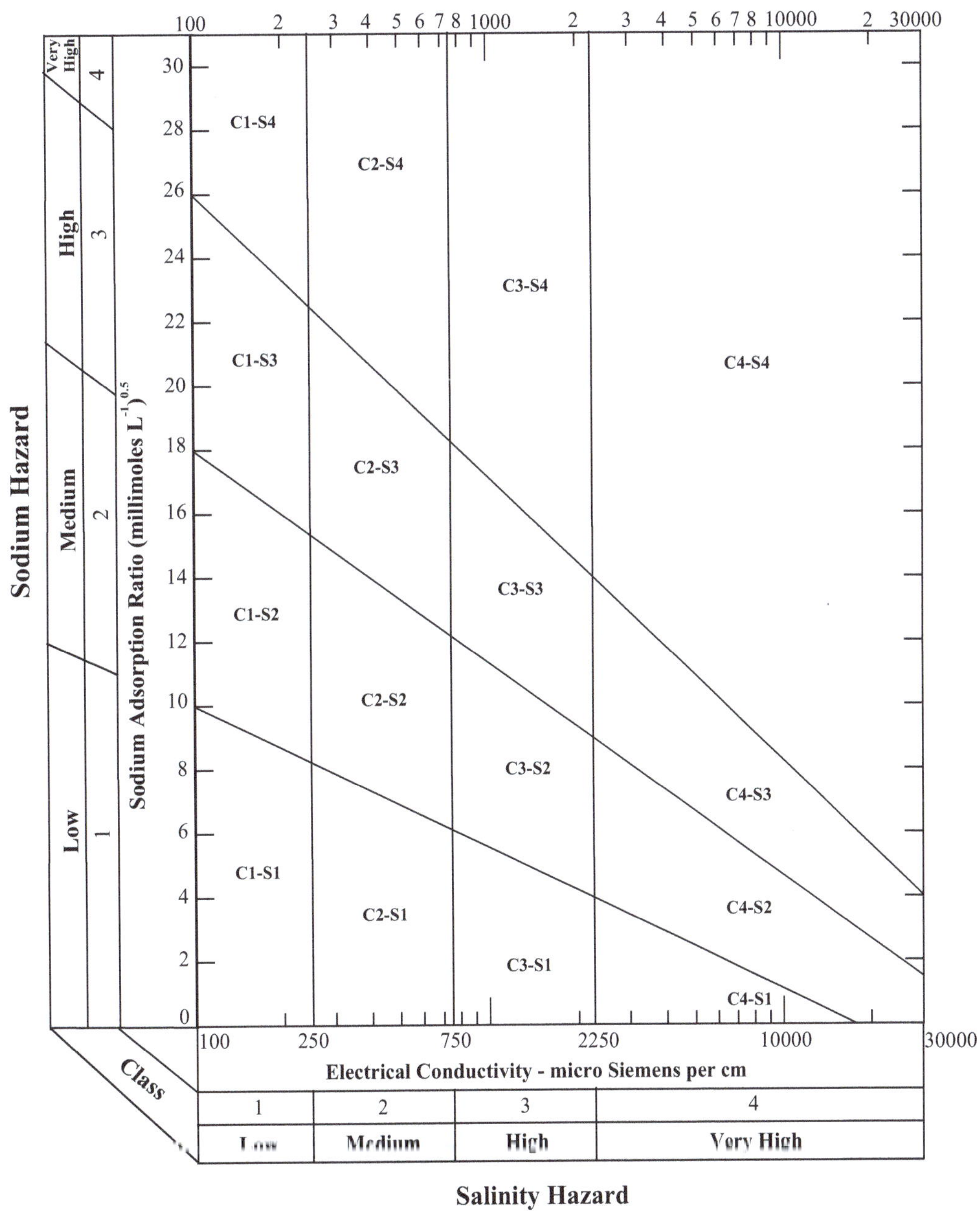

Fig. 5.1 Diagram for the classification of irrigation waters (USSL Staff 1954; modified by Shahid and Mahmoudi 2014)

2.2 Sodium Hazard

The sodium hazard of irrigation water is expressed as the 'sodium adsorption ratio (SAR)'. Although sodium contributes directly to the total salinity and may also be toxic to sensitive crops, such as fruit trees, the main problem with a high sodium concentration is its effect on the physical properties of soil (soil structure degradation). It is, thus, recommended to avoid using water with an SAR value greater than 10 (mmoles $l^{-1})^{0.5}$, if the water will be the only source of irrigation for long periods.

This recommendation holds even if the total salt content is relatively low. For example, if the soil contains an appreciable amount of gypsum, SAR value of 10 (mmoles $l^{-1})^{0.5}$ can be exceeded. The gypsum content of the soil should, thus, be determined.

Continued use of water with a high SAR value leads to a breakdown in the physical structure of the soil – a situation caused by excessive amounts of adsorbed sodium on soil colloids. This breakdown in the soil physical structure, results in the dispersion of soil clay and that causes the soil to become hard and compact when dry, and increasingly impervious to water penetration (due to dispersion and swelling) when wet. Fine textured soils, those high in clay, are especially subject to this action. When the concentration of sodium becomes excessive (in proportion to calcium plus magnesium), the soil is said to be sodic. If calcium and magnesium are the predominant cations adsorbed onto the soil exchange complex, the soil can be easily tilled and will have a readily permeable granular structure.

The permissible value of the SAR is a function of salinity. High salinity levels reduce swelling and aggregate breakdown (dispersion), thus promoting water penetration. A high proportion of sodium, however, produces the opposite effect.

Regardless of the sodium content, water with an electrical conductivity (EC) less than about 200 μS cm^{-1} causes degradation of the soil structure, promotes soil crusting and reduces water penetration. Rainfall is the prime example of low salinity water and rain water will reduce the penetration of water applied subsequently into soils. It is, thus, important that both the salinity and the sodium adsorption ratio of the applied water be considered when assessing the potential effects of water quality on water penetration into soils.

2.3 Carbonates and Bicarbonates Concentration

Waters high in carbonates (CO_3^{2-}) and bicarbonates (HCO_3^-) will tend to precipitate calcium carbonate ($CaCO_3$) and magnesium carbonate ($MgCO_3$), when the soil solution becomes concentrated through evapotranspiration. This means that the SAR value will increase, and the relative proportion of sodium ions will become greater. This situation, in turn, will increase the sodium hazard of the soil-water to a level greater than indicated by the SAR value.

2.4 *Specific Ion Effects (Toxic Elements)*

In addition to salinity and sodium hazards, certain crops may be sensitive to the presence of moderate to high concentrations of specific ions in the irrigation waters or soil solution. Many trace elements are toxic to plants at very low concentrations. Both soil and water testing can help to discover any constituents that might be toxic. Direct toxicity to crops may result from some specific chemical elements in irrigation water, e.g. boron, chloride, and sodium are potentially toxic to plants. The actual concentration of an element in water that will cause toxic symptoms varies, depending on the crop.

When an element is added to the soil through irrigation, it may be inactivated by chemical reactions. Alternatively, it may buildup in the soil until it reaches a toxic level. An element at a given concentration in water may be immediately toxic to a crop. Or, it may require a number of years to accumulate in the soil before it becoming toxic.

2.4.1 Sodium Toxicity

Sodium toxicity can occur in the form of leaf burn, leaf scorch and dead tissues running along the outside edges of leaves. In contrast, Cl^- toxicity is often seen at the extreme leaf tip. In tree crops, a sodium concentration (in excess of 0.25–0.5%) in the leaf tissue is often considered to be a toxic level of sodium. Correct diagnoses can be made from soil, water and plant tissue analysis.

Three levels of exchangeable sodium percentage (ESP) (FAO-UNESCO 1973; Pearson 1960; Abrol 1982), which correspond to three tolerance levels, are defined as: sensitive (ESP < 15), semi-tolerant (ESP 15–40) and tolerant (ESP > 40). The crops/plants listed as sensitive include, among others, beans, maize, peas, orange, peach, mung bean, mash, lentil, gram and cowpea. Semi-tolerant plants include carrot, clover, lettuce, berseem, oat, onion, radish, rye, sorghum, spinach, tomato, and tolerant plants include alfalfa, barley, beet, Rhoades grass and Karnal (Kallar) grass.

2.4.2 Boron Toxicity

Boron is essential to the normal growth of all plants, but the amount required is low. If it exceeds a certain level of tolerance depending on the crop, then boron may cause injury. The range between deficiency and toxicity of boron for many crops is narrow.

Table 5.2 Effects of boron (B) concentration in irrigation water on crops (Follett and Soltanpour 2002; Bauder et al. 2011)

Boron concentration (ppm)	Effect on crops
< 0.5	Satisfactory for all crops
0.5–1.0	Satisfactory for most crops
1.0–2.0	Satisfactory for semi-tolerant crops
2.0–4.0	Satisfactory for tolerant crops only

In order to sustain an adequate supply of boron to the plant at least 0.02 ppm of boron in the irrigation water may be required. However, to avoid toxicity, boron levels in irrigation water should, ideally, be lower than 0.3 ppm. Higher concentrations of boron will likely require that the intended crop type must first be evaluated with respect to its boron tolerance. Although boron toxicity is not a problem in most areas, it can be an important irrigation water quality parameter. Interestingly, plants grown in soils high in lime may tolerate higher levels of boron than those grown in non-calcareous soils.

Boron is weakly adsorbed by soils. Thus, its actual root-zone concentration may not vary in direct proportion to the degree that boron sourced from the irrigation water has been concentrated in the plant during growth. Symptoms of boron injury may include characteristic leaf 'burning', chlorosis and necrosis, although some boron sensitive species do not develop obvious symptoms. Boron toxicity symptoms first appear on older leaves as yellowing, spotting, or drying of leaf tissues at the tips and edges. The drying and chlorosis often progresses toward the center of the leaf, between the veins as boron accumulates over time (Ayers and Westcot 1985).

Irrigation water with boron >1.0 ppm may cause toxicity in boron sensitive crops. Table 5.2 describes the effects of a range of boron concentrations in irrigation water on crops (Bauder et al. 2011). The relative tolerance of plants to boron is shown in Table 5.3.

Boron levels that have developed in the soil water (saturation extract of soils) through irrigation can have a range of effects on crop yields. Wilcox (1960) presented three classes of crops with regard to boron toxicity: tolerant (2–4 ppm), semi-tolerant (1–2 ppm), and sensitive (0.3–1 ppm). Fruit crops are among the most boron sensitive, and yields of citrus and some stone fruit species are decreased by boron even at soil solution concentrations less than 0.5 ppm.

2.4.3 Chloride Toxicity

The most common crop toxicity is caused by chlorides in irrigation water. The chloride (Cl^-) anion occurs in all waters; chlorides are soluble and leach readily to drainage water. Chlorides are necessary for plant growth, though in high

Table 5.3 Relative tolerance[a] of plants to Boron concentration (ppm) in irrigation water (cf. Ludwick et al. 1990; Ayers and Westcot 1985)

Very sensitive < 0.5 ppm	Sensitive 0.5–0.75 ppm	Less sensitive 0.75–1.0 ppm	Moderately sensitive 1.0–2.0 ppm	Moderately tolerant 2.0–4.0 ppm	Tolerant 4.0–6.0 ppm	Very tolerant > 6.0 ppm
Lemon	Avocado	Garlic	Pepper, red	Lettuce	Tomato	Cotton
Blackberry	Grapefruit	Sweet potato	Pea	Cabbage	Parsley	Asparagus
	Orange	Sunflower	Carrot	Celery	Beet, red	
	Apricot	Bean	Radish	Turnip		
	Peach	Sesame	Potato	Oats		
	Cherry	Strawberry	Cucumber	Corn		
	Plum	Bean, kidney		Clover		
	Grape	Peanut		Squash		
	Walnut			Muskmelon		
	Onion					

Adapted from 'Salt Tolerance of Plants' (Maas 1987), In: CRC Handbook of Plant Science in Agriculture

[a]Maximum concentrations tolerated in soil-water or saturation extract without yield or vegetative growth reduction. Boron tolerance varies depending upon climate, soil conditions and crop varieties. Maximum concentrations in the irrigation water are approximately equal to these values or slightly less

Table 5.4 Chloride (Cl^-) levels of irrigation waters and their effects on crops (cf. Ludwick et al. 1990; Bauder et al. 2011)

Cl^- concentration		
meq l^{-1}	ppm	Effect on crops
< 2	< 70	Generally safe for all plants
2–4	70–140	Sensitive plants usually show slight to moderate injury
4–10	141–350	Moderately tolerant plants usually show slight to substantial injury
> 10	> 350	Can cause severe problems

concentrations they can inhibit plant growth, and can be highly toxic to some plant species. Water must, thus, be analyzed for Cl^- concentration when assessing water quality. Table 5.4 shows Cl^- levels in irrigation water and the effects of Cl^- on crops. In sensitive crops, symptoms occur when Cl^- levels accumulate in leaves (0.3–1.0% on a dry weight basis). Ayers and Westcot (1985) reported that Cl^- toxicity on plants appears first at the leaf tips (which is a very common symptom for chloride toxicity), and progresses from the leaf tip back along the edges as severity of the toxic effect increases. Excessive necrosis is often accompanied by early leaf drop or even total plant defoliation.

3 Classification of Irrigation Water

Shahid and Mahmoudi (2014) have modified the widely used USSL Staff (1954) salinity and sodium classification diagram for irrigation water (Fig. 5.1). This modified diagram is based on the EC (expressed in micro Siemens per cm – μS cm $^{-1}$) and the sodium adsorption ratio (SAR).

How to Use the Diagram?

The SAR as shown on y-axis (Fig. 5.1) can be calculated by using the following formula:

$$SAR = \frac{Na^+}{\sqrt{\frac{1}{2}\left(Ca^{2+} + Mg^{2+}\right)}}$$

Where, the concentrations of Na^+, Ca^{2+} and Mg^{2+} are expressed as milli equivalents per liter (meq l^{-1}). The values of the electrical conductivity given on the x-axis are expressed in micro Siemens per cm (μS cm^{-1}). The position of the SAR and EC points determines the quality class assigned to the water.

4 Analysis of Irrigation Water

4.1 Chemical Analyses

The ultimate in water quality data for appraisal of salinity and sodicity includes complete analyses for all major cations and anions for both irrigation and drainage waters. Major cations normally include Na^+, K^+, Ca^{2+} and Mg^{2+}. Major anions normally include CO_3^{2-}, HCO_3^-, Cl^- and also SO_4^{2-} (though see discussion below with regard to sulfate anion measurement).

When complete analyses are provided, it is possible to apply some simple tests for data consistency. For high quality water analysis, the sum of the cations in meq l^{-1} should be approximately equal to the sum of anions in meq l^{-1}. If the values are exactly equal, however, for several water samples, this suggests that some constituents have been estimated by 'difference'. For example, recent analyses of sulfate have commonly been determined by difference because of the general unavailability of a rapid and convenient analytical procedure for measuring sulfate (Bresler et al. 1982). The SO_4^{2-} estimation is based on the difference between total soluble cations and the sum of CO_3^{2-}, HCO_3^-, and Cl^-. In fact, sulfate is not a water constituent used to measure or determine either of SAR or Residual Sodium Carbonates (RSC). Thus, sulfate measurement currently has no assigned role in water quality assessment.

The data from above measurements are, thus, used to calculate the SAR in order to assess the sodicity hazard of the irrigation water, e.g. by use of Fig. 5.1 to obtain the water's sodicity (S) class. The EC, expressed in $\mu S\ cm^{-1}$, will then be used to obtain the conductivity (C) class of salinity. In addition, Residual Sodium Carbonate (RSC) can also be measured. These measurements are briefly described below.

4.1.1 EC and Total Salt Concentration

The most important water quality parameter from the standpoint of salinity is the total concentration of dissolved salts. It is different from 'total dissolved solids (TDS)', a term which carries some ambiguity. The measurement of TDS is much more tedious than measuring the EC – which is the preferred measure of salinity (Bresler et al. 1982). A simple meter is used to measure the electrical conductivity (EC) of both irrigation and drainage waters. Total salt concentration can then be obtained by using the following relationship for water having EC values between 0.1 and 10 milli Siemens per cm ($mS\ m^{-1}$) or $dS\ m^{-1}$ (Bresler et al. 1982):

$$\text{Total cations or anions}\left(\text{meq}\,l^{-1}\right) = 10 \times \text{EC}\left(\text{mS cm}^{-1}\text{or dS m}^{-1}\right)$$

Thus, once the concentrations of total cations or anions are known, the sum of cations or anions represents concentration of total salts contained within any solution.

4.1.2 Sodium Adsorption Ratio (SAR)

The tendency of salt solution to produce excessive exchangeable sodium in a soil must also be considered. A useful index for predicting this tendency is the Sodium Adsorption Ratio (SAR).

An SAR less than 8 $(\text{mmoles } l^{-1})^{0.5}$ is considered to be a 'low sodium' water class, i.e. the use of the irrigation water with SAR less than 8 is rated as being safe with regard to causing sodicity. That said, the prolonged use of class 8 SAR water for irrigation, when water drainage and leaching is restricted, may cause soils to develop sodicity. The detrimental effect of SAR also depends on the EC value, and in Pakistan an SAR of 10 is considered safe level (Kinje 1993).

Adjusted SAR

The significance of SAR_{adj} is that under field conditions, and in normal conditions of irrigation management, the exchangeable sodium percentage (ESP) value in top soil is very nearly equal to the adjusted SAR, where pHc is calculated as the pH used in the Langelier Index of the irrigation water. Ayers and Westcot (1985) presented the term adjusted SAR (SAR_{adj}) as:

$$SAR_{adj} = SAR_{IW}[1 + (8.4 - pHc)]$$

The Langelier index is based on calculation of the pH which given water would achieve when in equilibrium with solid-phase calcium carbonates at average CO_2 values. This pH, when compared to the initial pH of the water, can be used to predict whether $CaCO_3$ should precipitate from or be dissolved by the waters as it passes through calcareous soil (Balba 1995). The pHc is the theoretical pH that water could have in equilibrium with $CaCO_3$.

4.1.3 Residual Sodium Carbonates (RSC)

There is another approach which is empirical in nature (Eaton 1950). It has been widely used to predict the additional sodium hazard which is associated with $CaCO_3$ and $MgCO_3$ precipitation, and involves a calculation of the residual sodium carbonates (RSC). This approach is based on the equation:

$$RSC\ \left(meq\ l^{-1}\right) = \left(CO_3^{2-} + HCO_3^-\right) - \left(Ca^{2+} + Mg^{2+}\right)$$

Where, all the concentrations are in meq l^{-1}. The ranges of RSC in meq l^{-1} with respect to water suitability for irrigation are shown in Table 5.5.

5 Conductivity Classes (USSL Staff 1954)

There are four salinity classes, low, medium, high and very high, as presented in Table 5.6.

5.1 Low Salinity Water (Salinity Class C1)

It can be used for irrigation of most crops on most soils with little likelihood that soil salinity will develop. Some leaching will be required for salinity Class C1 water, but

Table 5.5 Residual sodium carbonates (RSC) and suitability of water for irrigation (Eaton 1950; Wilcox et al. 1954)

RSC (meq l^{-1})	Suitability of water for irrigation
< 1.25	Safe
1.25–2.50	Marginal
> 2.5	Unsuitable

Table 5.6 Salinity classes of irrigation waters (USSL Staff 1954)

Salinity of irrigation water – EC (μS cm^{-1})	Salinity class	Salinity hazard
100–250	C1	Low
250–750	C2	Medium
750–2250	C3	High
> 2250	C4	Very high

this occurs under normal irrigation practices, except for soils with extremely low permeability.

5.2 *Medium Salinity Water (Salinity Class C2)*

It can be used if a moderate amount of leaching can occur. Plants with moderate salt tolerance can be grown in most cases without special practices for salinity control.

5.3 *High Salinity Water (Salinity Class C3)*

It cannot be used on soils which possess restricted drainage and, thus, poor leaching abilities. Even with adequate drainage, special management for salinity control may be required and plants with good salt tolerance should always be selected.

5.4 *Very High Salinity Water (Salinity Class C4)*

It is not suitable for irrigation under ordinary conditions, but may be used occasionally under very special circumstances. Here, the soils must be permeable, drainage must be adequate to good and irrigation water must be applied in excess in order to provide considerable leaching. Only very salt tolerant crops should be selected.

6 Sodicity Classes (USSL Staff 1954)

The classification of irrigation waters with respect to sodium adsorption ratio (SAR) is based primarily on the effects which exchangeable sodium accumulation has on the physical conditions of the soil. However, it should be kept in mind that sodium sensitive plants may still suffer injury (as a result of sodium accumulation in plant tissues) even when exchangeable sodium values in soil-water are too low to bring about a deterioration of the physical condition of the soil.

6.1 Low Sodium Water (Sodicity Class S1)

It can be used for irrigation on almost all soils with little danger of the soil developing harmful levels of exchangeable sodium. However, sodium sensitive crops such as stone fruit trees and avocados may accumulate injurious concentrations of sodium.

6.2 Medium Sodium Water (Sodicity Class S2)

It will present an appreciable sodium hazard in fine textured soils which have high cation exchange capacity, especially under low leaching conditions, unless gypsum is present in the soil. Sodicity class S2 water may be used in coarse textured or organic soils with good permeability.

6.3 High Sodium Water (Sodicity Class S3)

It may produce harmful levels of exchangeable sodium in most soils. Its use will require special soil management methods, good drainage, a high leaching ability and high organic matter conditions. Gypsiferous soils, however, may not develop harmful levels of exchangeable sodium from such waters. Management methods may require use of chemical amendments which encourage the replacement of exchangeable sodium. That said, use of those amendments may not be feasible with waters of very high salinity.

6.4 Very High Sodium Water (Sodicity Class S4)

It is generally unsatisfactory for irrigation purposes except at low and perhaps medium salinity. Specifically, where the soil water solution is rich in calcium or the use of gypsum or other soil amendments may make the use of sodicity class S4 irrigation water feasible. Irrigation water sodicity classes and their hazards are given in Table 5.7.

Table 5.7 Sodicity classes of irrigation water (USSL Staff 1954)

SAR of irrigation water (mmoles $l^{-1})^{0.5}$	Sodicity class	Sodicity hazard
< 10	S1	Low
10–18	S2	Medium
18–26	S3	High
> 26	S4	Very high

Sometimes the irrigation water may dissolve sufficient calcium from calcareous soils to decrease the sodium hazard appreciably, and this should be taken into account using salinity class C1 – sodicity class S3 and salinity class C1 – sodicity class S4 irrigation waters. For calcareous soils with high pH values, or for non-calcareous soils, the sodium status of irrigation water in salinity class C1 – sodicity class S3, salinity class C1 – sodicity class S4, and salinity class C2 – sodicity class S4 may be improved by the addition of gypsum through lining of irrigation channels with gypsum stones or the sodium hazard may be countered by applying gypsum to the soil periodically. This is especially applicable when salinity class C2 – sodicity class S3 and salinity class C3 – sodicity class S2 irrigation water is used.

7 Improvement of Irrigation Water Quality

There are a number of ways to improve water quality, with regard to salinity and sodicity hazards, prior to using for irrigation purposes. Most commonly used practices are described below.

7.1 Blending Water

The saline/brackish water quality can be improved if an alternate source of good quality water is available. The desired water salinity level, depending upon the crop to be irrigated, can be derived by a standard calculation procedure.

Example

A blend is made with 50% fresh water (EC 0.25 dS m^{-1}) with 50% brackish water (EC 3.9 dS m^{-1}). The resulting EC of the blended water would be:

$$\begin{aligned}&\text{EC(blended water)} = (\text{EC of fresh water} \times \text{mixing ratio}) +\\&(\text{EC of brackish water} \times \text{mixing ratio}) = (0.25 \times 0.50) + (3.90 \times 0.50)\\&= 0.125 + 1.95 = 2.075 \text{dS m}^{-1}\end{aligned}$$

7.2 *Blending Water to Achieve a Desired Salinity*

The desired water salinity can be achieved (by mixing two waters of known salinity) to irrigate a specific crop based on the threshold salinity. In this case, it is necessary to know what ratio of the two waters will be used to achieve the desired salinity.

Example
A blend is to be made of two waters, fresh (0.25 dS m^{-1}) with brackish (20 dS m^{-1}). Thus, we need to know 'in what ratio these two waters are to be mixed' to achieve a desired resultant water salinity of 8 dS m^{-1}.

Let us assume that we need to develop a final volume of 2 liters of the resultant water with a salinity of 8 dS m^{-1}.

$$\text{A standard formula can be used} : C1V1 = C2V2$$

Where,

$C1 = 20$ dS m^{-1}
$V1$ = unknown volume of the brackish water
$C2 = 7.75$ dS m^{-1} or desired water salinity ($8–0.25 = 7.75$)
$V2 = 2$ liters or 2000 ml of desired final volume

Using the formula,

$$C1V1 = C2V2$$
$$20 \times V1 = 7.75 \times 2000 \text{ ml}$$
$$V1 = (7.75 \times 2000 \text{ ml})/20 = 775 \text{ ml}$$

Thus, 775 ml of the brackish water will be required to raise EC of the fresh water from 0.25 to 8 dS m^{-1}. The resulting blending ratio will be (1:2.58, i.e. the ratio of brackish water added to fresh water).

8 Water Sodicity Mitigation

Water sodicity can be mitigated through the judicious use of calcium-containing amendments such as gypsum ($CaSO_4.2H_2O$). Relative to other amendments, gypsum is cheap and easy to handle, and by far the most suitable amendment to bring down irrigation water sodicity (the ratio of sodium to calcium + magnesium). The quantity of gypsum needed for adding to irrigation water depends upon the quality of water (RSC and SAR levels) and the quantity of water required for irrigation during the growing season of the crop.

8.1 *Gypsum Requirement Using the Residual Sodium Carbonates (RSC) Concept*

Example 1
Irrigation water has an RSC 8.5 meq l^{-1} and it needs to be reduced to 2.5 meq l^{-1}. The water required for irrigation is 800 mm per hectare for the complete growing period of the sorghum crop. How much gypsum will be required for adding to the water that is needed to irrigate one hectare, that water having the desired RSC of 2.5 meq l^{-1}?

- 1equivalent per liter of Na^+ will require 1 equivalent per liter of Ca^{2+} which is equal to 86.06 grams of gypsum per liter of solution
- Therefore, 1 meq l^{-1} of Na^+ will require 1 meq l^{-1} of Ca^{2+} which is equal to 0.08606 grams of gypsum per liter of solution
- Thus, 6 meq l^{-1} of Na^+ will require 6 meq l^{-1} of Ca^{2+} which is equal to 0.51636 grams of gypsum per liter of solution
- Total water required to irrigate one hectare of sorghum crop = 800 mm × 10 = 8000 M^3 (Where, 1 mm of water in 1 hectare is equal to 10 M^3)
- 8000 M^3 of water is equal to 8000 × 1000 = 8,000,000 liters of irrigation water across the entire growing season
- Total gypsum requirement = 8,000,000 × 0.51636 = 4.13 metric tons of 100% pure gypsum
- If the gypsum purity is 70%, then 5.90 tons of gypsum will be required to neutralize 6 meq l^{-1} of Na^+ in 8 million liters of irrigation water

To amend the water RSC, it is best to place the gypsum in the water channels. Then, the flowing irrigation water will dissolve the gypsum, reducing the $Na^+:(Ca^{2+} + Mg^{2+})$ ratio prior to entering the agricultural field.

Example 2
A farmer is using saline water with an EC of 3 dS m^{-1} for irrigating a sorghum crop. He is facing problems with irrigation water infiltrating into his field soil and has decided to use gypsum. A laboratory analysis has shown that he needs an increase of 5 meq l^{-1} of calcium in the irrigation water. How much gypsum would be required to irrigate one-hectare area with a crop water requirement for the entire growing period as 800 mm?

- EC of water = 3 dS m^{-1}
- Cropped area = 1 ha
- Gypsum purity = 70%

Total water requirement = 800 mm × 10 = 8000 M^3 = 8,000,000 liters.

- 1 meq l^{-1} of Na^+ will require 1 meq l^{-1} of Ca^{2+} which is equal to 0.08606 grams of gypsum per liter of solution.

Table 5.8 The chemical analyses of well water

Water	EC dS m^{-1}	Ion concentrations (meq l^{-1})								SAR (mmoles $l^{-1})^{0.5}$
		Na^+	K^+	Ca^{2+}	Mg^{2+}	CO_3^-	HCO_3^{2-}	Cl^-	SO_4^{2-}	
Well water	4	25	2	7	6	0	0	20	20	9.805
Resultant water	1	6.25	0.5	1.75	1.5	0	0	5	5	4.903

- 5 meq l^{-1} of Na^+ will require 5 meq l^{-1} of Ca^{2+} which is equal to 0.4303 grams of gypsum per liter of solution.
- Total water required to irrigate one hectare of sorghum crop = 800 mm or 8000 M^3
- 8000 M^3 of water is equal to 8000 × 1000 = 8,000,000 liters.
- Total gypsum requirement = 8,000,000 × 0.4303 = 3.44 metric tons of 100% pure gypsum
- If gypsum purity is 70%, then 4.92 metric tons of gypsum will be required to neutralize 5 meq l^{-1} of Na^+ in 8 million liters of water.

Thus, 4.91 tons of gypsum of about 10 mesh size (2 mm) will be required for the irrigation water application across the entire growing season.

8.2 *Determining the SAR of Blended Water to Be Used for Irrigation*

Example 1

Water from a well has the composition (Table 5.8) and this well water will be diluted in a1:3 ratio with desalinated water. What will be the resultant SAR of the blended water? Assume that the desalinated water has negligible EC and Na^+, Ca^{2+}, Mg^{2+} contents.

After blending with a ratio of 1:3 (well water:desalinated water), the SAR of the resultant blended water is reduced to half. However, it should be noted that the EC is reduced to one-quarter of the well water. Therefore, care should be taken to understand such conversions.

Example 2

A canal water (EC = 1.0 dS m^{-1}) source is available to irrigate a crop. However, the volume of water is insufficient. The farmer has decided to blend well water with a ratio of 20% well water (5 dS m^{-1}) with 80% of canal water (1 dS m^{-1}). What will be the SAR of the resultant water? Following are the water analyses of canal, well and blend waters (Table 5.9).

Table 5.9 The chemical analyses of the canal, well and the resultant (blended) waters

Water	EC ($dS\ m^{-1}$)	Ion concentrations ($meq\ l^{-1}$)								SAR ($(mmoles\ l^{-1})^{0.5}$)
		Na^+	K^+	Ca^{2+}	Mg^{2+}	CO_3^{2-}	HCO_3^-	Cl^-	SO_4^{2-}	
Canal water	1.0	6.25	0.5	1.75	1.5	0	0	5.0	5.0	4.903
Well water	5.0	32.0	2.5	9.0	8.0	0	0	25.0	25.0	10.98
Blended water	1.8	11.4	0.9	3.2	2.8	0	0	9.0	9.0	6.58

Composition of blended water:

EC	$= (1.0 \times 0.8) + (5.0 \times 0.20)$	$= 0.8 + 1.0$	$= 1.8\ dS\ m^{-1}$
Ca^{2+}	$= (1.75 \times 0.8) + (9.0 \times 0.2)$	$= 1.4 + 1.8$	$= 3.2\ meq\ l^{-1}$
Mg^{2+}	$= (1.5 \times 0.8) + (8 \times 0.2)$	$= 1.2 + 1.6$	$= 2.8\ meq\ l^{-1}$
Na^+	$= (6.25 \times 0.8) + (32.0 \times 0.2)$	$= 5.0 + 6.4$	$= 11.4\ meq\ l^{-1}$
K^+	$= (0.5 \times 0.80) + (2.5 \times 0.20)$	$= 0.4 + 0.5$	$= 0.9\ meq\ l^{-1}$
Cl^-	$= (5.0 \times 0.80) + (25.0 \times 0.2)$	$= 4.0 + 5.0$	$= 9\ meq\ l^{-1}$
SO_4^{2-}	$= (5.0 \times 0.80) + (25.0 \times 0.2)$	$= 4.0 + 5.0$	$= 9\ meq\ l^{-1}$
SAR	$= Na^+/[(Ca^{2+} + Mg^{2+})/2]^{0.5}$	$= 11.4/[(3.2 + 2.8)/2]^{0.5}$	$= 6.58\ (mmoles\ l^{-1})^{0.5}$

Blending should, thus, be done with an objective. If the objective is to reduce SAR, but with the condition that adequate canal/fresh water is not available to irrigate the crop, then blending is desirable. If, however, a sufficient volume of canal water is available, then simply replacing well water with the canal's fresh water for irrigation is a good option. Other farm conditions must also be considered, e.g. infiltration problems due to high SAR. Addition of gypsum as described above should also be considered.

9 Cyclic Use of Water

Where fresh water is also available, but not sufficient to offset the full water requirement of the crop, there is always a need to find alternate source of water, which is usually the groundwater and is often saline or saline-sodic. Under such conditions, it is recommended to use fresh water at early stage of crop when the young seedlings are not able to tolerate high salinity level. Once the seedlings are well established, at this stage there are two options to use these waters: (i) to use saline water for some time and then leach the salts with fresh water, and (ii) use saline water first and then use fresh water (cyclic use) to irrigate the crop. This way both fresh and saline waters are used.

References

Abrol IP (1982) Technology of chemical, physical and biological amelioration of deteriorated soils. Panel of experts meeting on amelioration and development of deteriorated soils in Egypt 2–6 May 1982, Cairo, pp 1–6

Ayers RS, Westcot DW (1985) Water quality for agriculture. FAO irrigation and drainage paper 29 rev 1. Food and agriculture organization of the United Nations, Rome, Italy, 174 pp
Balba AM (1995) Management of problem soils in arid ecosystems. CRC/Lewis Publishers, Boca Raton, 250 pp
Bauder TA, Waskom RM, Sutherland PL, Davis JG (2011) Irrigation water quality criteria. Colorado State University Extension Publication, Crop series/irrigation. Fact sheet no. 0.506, 4 pp
Bresler E, McNeal BL, Carter DL (1982) Saline and sodic soils. Principles-dynamics-modeling. Advanced Series in Agricultural Sciences 10. Springer-Verlag, Berlin/Heidelberg/New York, 236 pp
Eaton FM (1950) Significance of carbonates in irrigation waters. Soil Sci 69:123–133
FAO/UNESCO (1973) Irrigation, drainage and salinity. An International source book. Unesco/FAO, Hutchinson & Co (Publishers) Ltd, London, 510 pp
Follett RH, Soltanpour PN (2002) Irrigation water quality criteria. Colorado State University Publication No. 0.506
Kinje JW (1993) Environmentally sound water management: Irrigation and the environment. Proceedings of the International Symposium on Environmental Assessment and Management of Irrigation and Drainage Projects for Sustained Agricultural Growth, 24–28 October 1993, Lahore, Pakistan, pp 14–44
Ludwick AE, Campbell KB, Johnson RD, McClain LJ, Millaway RM, Purcell SL, Phillips IL, Rush DW, Waters JA (eds) (1990) Water and plant growth. In: Western Fertilizer Handbook – horticulture Edition, Interstate Publishers Inc, Illinois, pp 15–43
Maas EV (1987) Salt tolerance of plants. In: Christie BR (ed) Handbook of plant science in agriculture. CRC Press, Boca Raton, pp 57–75
Pearson GA (1960) Tolerance of crops to exchangeable sodium. USDA Information Bulletin No 216, 4 pp
Shahid SA (2004) Irrigation water quality manual. ERWDA Soils Bulletin No 2, 29 pp
Shahid SA, Mahmoudi H (2014) National strategy to improve plant and animal production in the United Arab Emirates. Soil and water resources Annexes
USSL Staff (1954) Diagnosis and improvement of saline and alkali soils. USDA Handbook No 60. Washington DC, USA 160 pp
Wilcox LV (1960) Boron injury to plants. USDA Bulletin No 211, 7 pp
Wilcox LV, Blair GY, Bower CA (1954) Effect of bicarbonate on suitability of water for irrigation. Soil Sci 77:259–266

Chapter 6
The Role of Nuclear Techniques in Biosaline Agriculture

Mohammad Zaman, Shabbir A. Shahid, and Lee Heng

Abstract The major constraints under Saline Agriculture are the availability of essential nutrients and water to the plant which are adversely affected by excessive salts in the soil solution. Among the essential plant nutrients, N plays a key role in plant growth and productivity. Nuclear and isotopic techniques (also called nuclear-based techniques) are a complement to, not a substitute for, non-nuclear conventional techniques. Nuclear-based techniques, however, do have several advantages over conventional techniques by providing unique, precise and quantitative data on soil nutrient and soil moisture pools and fluxes in the soil-plant-water and atmosphere systems. Isotopic techniques provide useful information in assessing soil-water-nutrient management which can be tailored to specific agroecosystems for managing soil salinity. For example, ^{15}N stable isotopic techniques can be used to measure rates of the various N transformation processes in soil-plant-water and atmosphere systems, such as N mineralization-immobilization, nitrification, biological N_2 fixation, N use efficiency, and microbial sources of production of nitrous oxide (N_2O), a greenhouse and ozone depleting gas, in soil. The use of oxygen-18, hydrogen-2 (deuterium) and other isotopes is an integral part of agricultural water management, allowing the identification of water sources and the tracking of water movement and pathways within agricultural landscapes as influenced by different irrigation technologies, cropping systems and farming practices. It also helps in the understanding of plant water use, quantifying crop transpiration and soil evaporation and allows us to devise strategies to improve crop production, reduce unproductive water losses and prevent land and water degradation.

Keywords Isotopic and nuclear techniques · N-15 · Oxygen-18 · Hydrogen-2 · Salinity

M. Zaman et al., *Guideline for Salinity Assessment, Mitigation and Adaptation Using Nuclear and Related Techniques*, https://doi.org/10.1007/978-3-319-96190-3_6

1 Introduction

Among the numerous abiotic and biotic stresses that affect plant productivity worldwide, soil water stress (drought) is the most common growth limiting factor in arid and semi-arid regions (Saranga et al. 2001), followed closely by salt stress (Pessarakli 1991). Development of a sustainable agriculture will require the combined use of soil, nutrient, and water management strategies that enhance crop productivity, while at the same time reducing abiotic and biotic stresses. To reach a truly sustainable agriculture, new 'climate smart' agricultural practices will need to be developed and adopted by the end users. These climate smart practices include both management strategies and specific technologies, ones which enhance crop productivity, environmental sustainability and wise use (conservation) of agro-ecosystems.

The Soil and Water Management & Crop Nutrition (SWMCN) subprogram of the Joint Food and Agriculture Organization (FAO) and International Atomic Energy Agency (IAEA)'s Division of Nuclear Applications in Food and Agriculture, has developed a wide range of nuclear and isotopic techniques to enhance nutrient and water use efficiencies, increase biological N fixation through the capture of atmospheric di-nitrogen (N_2) and carbon (C) storage in salt affected soil.

2 Background Information on Isotopes

The number of protons plus neutrons present in the nucleus of an atom is called the atomic weight, while the number of protons (or electrons – which is always equal) is known as atomic number. Isotopes are defined as atoms of the same atomic number but differing atomic weight. For example, nitrogen (N) has one isotope (^{15}N), which has the same number of protons (7) as ^{14}N, but one extra neutron. This gives it (^{15}N) a different atomic weight ($7 + 8 = 15$).

Isotopes may exist in both stable and unstable (radioactive) forms, depending on the stability of the nucleus in an atom. For example, the sulfur (S) consists of 5 isotopes (^{32}S, ^{33}S, ^{34}S, ^{35}S and ^{36}S); one of which (^{35}S) is a radioactive beta emitter, while the other four (^{32}S, ^{33}S, ^{34}S and ^{36}S) are stable. Thus, a radioactive isotope is an atom with an unstable nucleus which spontaneously emits radiation (alpha or beta particles and/or gamma electromagnetic rays). The non-stability occurs because the ratio of neutrons to protons in a nucleus lies outside the belt of stability (i.e., outside a particular number due to an excess of either protons or neutrons), which varies with each atom. In contrast, a stable isotope is an atom with a stable nucleus (i.e., the ratio of neutrons to protons in the nucleus of an atom is within the belt of stability), and hence, it does not spontaneously emit any radiation (Nguyen et al. 2011). Stable isotopes exist in light and heavy forms with heavy isotopes having a higher atomic weight than light isotopes (Table 6.1).

Table 6.1 Average abundances of stable isotopes (% abundance in brackets) of some of the major elements commonly occurring in agro-ecosystems

Element	Heavy isotope	Light isotope
Carbon	^{13}C (1.108%)	^{12}C (98.892%)
Hydrogen	^{2}H (0.0156%)	^{1}H (99.984%)
Nitrogen	^{15}N (0.366%)	^{14}N (99.634%)
Oxygen	^{18}O (0.204%)	^{16}O (99.759%)
	^{17}O (0.037%)	
Sulfur	^{33}S (0.76%)	^{32}S (95.02%)
	^{34}S (4.22%)	
	^{36}S (0.02%)	

The quantity of a stable isotope is measured by an Elemental Analyser coupled to an Isotope Ratio Mass Spectrometer (IRMS). Thus, a sample of soil or biological material is combusted into a gas, which is fed into a mass spectrometer, where the ratio of the stable isotopes of interest (e.g., $^{13}C/^{12}C$, $^{2}H/^{1}H$, $^{15}N/^{14}N$, $^{18}O/^{16}O$,$^{33}S/^{32}S$) is determined.

Radioactive isotopes (radioisotopes) are measured by their rate of 'decay', e.g. liquid scintillation counters are used for beta particle emitting radioactive isotopes, gamma spectrometers for gamma ray emitting radioactive isotopes and alpha spectrometers for alpha particle emitting radioactive isotopes. The international unit (SI) of activity decay is the Becquerel (Bq), which is equal to one disintegration per second (dps). The old unit commonly used was called the Curie, which is equivalent to 3.7×10^{10} dps or 3.7×10^{10} Bq (Nguyen et al. 2011).

3 Use of Nuclear and Isotopic Techniques in Biosaline Agriculture

Nuclear and isotopic techniques (also called nuclear-based techniques) are a complement to, not a substitute for, non-nuclear conventional techniques. Nuclear-based techniques, however, do have several advantages over conventional techniques by providing unique, precise and quantitative data on soil nutrient and soil moisture pools and fluxes in the soil-plant-water and atmosphere systems. Isotopic techniques provide useful information in assessing soil-water-nutrient management which can be tailored to specific agro-ecosystems for managing soil salinity. For example, ^{15}N stable isotopic techniques can be used to measure rates of the various N transformation processes in soil-plant-water and atmosphere systems, such as N mineralization-immobilization, nitrification, biological N_2 fixation, N use efficiency, and microbial sources of production of nitrous oxide (N_2O), a greenhouse and ozone depleting gas, in soil. Several nuclear and isotopic techniques are being employed in soil water management studies. The soil moisture neutron probe is ideal in field-scale rooting zone measurement of soil water, providing accurate data on the availability of water for determining crop water use and water use efficiency and for establishing optimal irrigation scheduling under different cropping systems especially under saline conditions.

The use of oxygen-18, hydrogen-2 (deuterium) and other isotopes is an integral part of agricultural water management, allowing the identification of water sources and the tracking of water movement and pathways within agricultural landscapes as influenced by different irrigation technologies, cropping systems and farming practices. It also helps in the understanding of plant water use, quantifying crop transpiration and soil evaporation and allows us to devise strategies to improve crop production, reduce unproductive water losses and prevent land and water degradation.

For details on the principles and applications of the various nuclear and isotopic techniques in soil, water and plant nutrient studies in agro-ecosystems, the readers are referred to the IAEA Training Manuals (IAEA 1990, 2001) and the review paper published by Nguyen et al. (2011). In below section, a stepwise protocol has been described to set up a field study to quantify fertilizer use efficiency of the added fertilizer.

4 The Use of Nitrogen-15 (^{15}N) to Study Fertilizer Use Efficiency

The major constraints under *Saline Agriculture* are the availability of essential nutrients and water to the plant which are adversely affected by excessive salts in the soil solution. Among the essential plant nutrients, N plays a key role in plant growth and productivity. To take up N from the soil solution, plants compete with a range of N removal processes/losses including immobilization, leaching, and gaseous emissions of N as ammonia (NH_3), nitrous oxide (N_2O), nitric oxide (NO) and molecular nitrogen (N_2) into the atmosphere. Because of these N losses, the N use efficiency (kg of dry matter produced per kg of N applied) or useful use of N by plant is invariably less than 50% of the applied N (Zaman et al. 2013a, b, 2014). The extent to which N is removed from soils, or made unavailable to plants by the above biogeochemical processes is of both economic and environmental importance.

Under saline conditions, the presence of excessive salts (especially Na^+) in the soil solution, coupled with a high soil pH, is likely to further increase the competition between N uptake by the plant and the soil N losses, thereby reducing crop productivity further. Quantifying N use efficiency and the sources of N losses enables researchers to develop ‘technology packages’ which can enhance N uptake and minimize N losses, thus allowing for sustainable crop productivity under saline conditions.

Fig. 6.1 A wheat trial set up on a flat soil

4.1 Setting Up Experimental Field Plots

In order to determine the N fertilizer use efficiency (NUE) of a wheat crop with a high degree of accuracy, a researcher shall set up a field trial on a relatively flat site with uniform fertility and slope so as to minimize background variations of soil nutrient levels, especially N and nutrients losses via surface runoff (Fig. 6.1).

Considering an experimental trial of N fertilizer applied at four rates: zero or control (T1), low (T2), middle (T3), and high (T4) of kg N per ha, with four individual replicate plots (each plot being 7 m × 7 m) for each of the four rates of N. (see schematic diagram below – Fig. 6.2).

A 'buffer zone' of 2 m wide on each of the four sides of the experimental site, with a 2 m wide strip between each of the individual replicate plots is especially important to prevent contamination of adjacent plots by N via surface runoff after heavy irrigation or rainfall, as well as lateral movement of N within the soil. The individual (replicate) field plots can be a range of sizes, depending on available land area, experimental design, farm resources (machinery) and most importantly available budget. Generally, a larger size for each individual replicate plot (e.g., 7 m long × 7 m wide) is considered as the best for minimizing edge effects (nutrient losses from the fertilized area to an un-fertilized area) on final crop yield, with each of four replicate plots being placed within four different treatment blocks.

- Prior to treatment application, four composite soil samples (each composite soil sample consist of ten soil cores from each experimental block) from 0–15 cm depth, shall be collected to analyze for key soil properties including, soil pH, ECe, Na^{+}, Ca^{2+}, Mg^{2+}, K^{+}, total N, total C, and Olsen P.

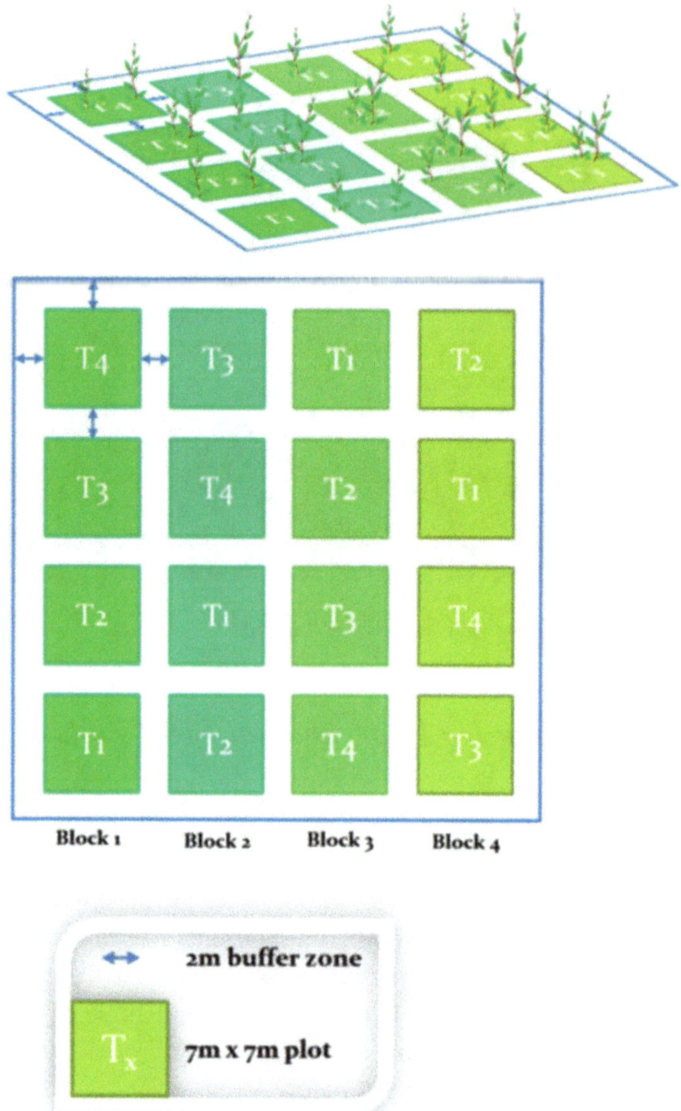

Fig. 6.2 A schematic diagram of experimental layout

- First apply any soil amendments such as gypsum, and other chemical fertilizers without N (P and K as recommended) and animal manure.
- Assuming 7 m × 7 m (49 m^2) replicated field plot receiving N-fertilizer in the form of granular urea (46%N) at rate of 80 kg N ha^{-1} in two split applications during wheat growth period, the amount of urea is calculated below:

$$\text{Rate of fertilizer application (kg per ha)} = \frac{100 \times \text{nutrient element required (kg per ha)}}{\%\text{nutrient element concentration in a fertilizer.}}$$

Example:

The amount of urea for the first application (40 kg N ha^{-1}) can be calculated as.

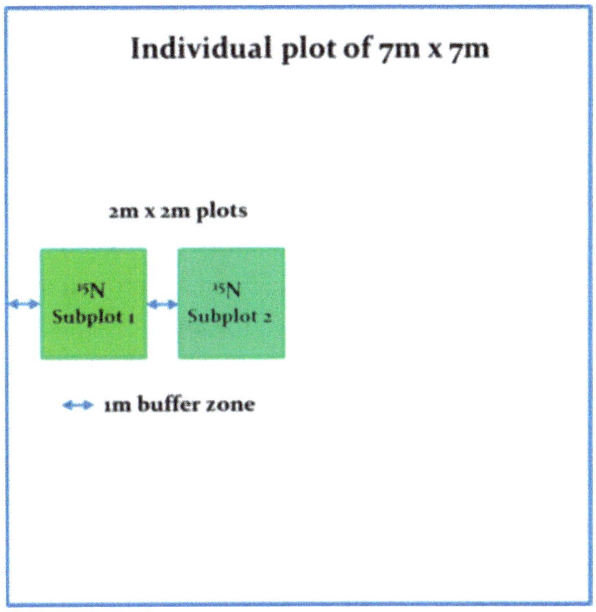

Fig. 6.3 Schematic diagram of the layout of the two sub-plots within a main plot, each with a 1 m buffer zone, each destined for ^{15}N-labeled fertilizer application

$$\text{kg of urea required per ha for the first application} = \frac{100 \times 40}{46} = 86.95\ \text{kg urea} \tag{6.1}$$

As mentioned below, during the N fertilization, one sub-plot (4 m^2) of ^{15}N labeled urea within the 49 m^2 replicated plot will not receive ordinary urea. This leaves 45 m^2 area (49 minus 4) which will receive ordinary urea. Thus at 40 kg N ha^{-1} rate, the amount of urea for 45 m^2 is calculated as:

$$\text{Amount of urea for } 45\ \text{m}^2 = \frac{86.95\ \text{kg urea}}{10\,000\ \text{m}^2} \times 45\ \text{m}^2 = 0.39\ \text{kg} \tag{6.2}$$

Where, 10,000 m^2 correspond to the land area of one hectare.

Setting up Sub-Plot for ^{15}N Labelled Fertilizer:

- For two split applications of ^{15}N-labelled urea, one shall set up two sub-plots, each of 2 m × 2 m (4 m^2), separated within the entire 49 m^2 larger replicated plot by a 1 m buffer zone, as shown below (Fig. 6.3). This 4 m^2 sub-plot will allow researcher to select a few wheat plants for ^{15}N analysis. The buffer zone will also help to minimize ^{15}N contamination from adjacent sub-plot.

  [Mark each sub-plot well to avoid any mistake of fertilizer application].

- To ensure that no ^{15}N-labeled fertilizer/residues are present from previous experiments, collect four soil cores (0–10 cm soil depth) from each of the two sub-plots, then combine them into one sample, and analyze for ^{15}N content. This will establish the initial ^{15}N level in the soil.

- Calculate the amount of ^{15}N-labeled fertilizer (using a maximum of 5 atom % excess) to add to each **4** m^2 sub-plot using Eqs. 6.1 and 6.2. The amount of ^{15}N-labeled urea at 40 kg N ha^{-1} for a 4 m^2 sub-plot comes out to be 34.78 gram.
- Please note that if N fertilizer is applied in a single application, this 5 atom % excess could be reduced to 3 atom % excess (please refer to the dilution procedure at the end of this section).
- Separate the first sub-plot for ^{15}N-labeled fertilizer by placing a temporary plastic sheet or any other similar material around the perimeter of the first sub-plot. Then, uniformly apply the required amount (0.39 kg) of ordinary urea to the entire (45 m ^{2}) of the larger main plot excluding the first sub-plot.
- After application of the ordinary urea, remove the plastic sheet around the first sub-plot, carefully weigh out the exact amount of ^{15}N-labeled fertilizer (34.78 gram) using Eq. 6.2, and apply ^{15}N-labeled urea evenly by hand to the first sub-plot. One shall be aware that ^{15}N-labeled fertilizer such as urea come as a fine particle therefore extreme care shall be taken while applying to ensure its even application. Fine sand of the same diameter or any other inert material shall be mixed with the ^{15}N-labelled urea to ensure even application. One shall also avoid ^{15}N labelled urea under windy conditions or when a heavy rainfall is expected. If irrigation water is available, it is important that the experimental plots are supplied with at least 10–20 mm of irrigation soon after N fertilizer application to move urea from surface into the soil to minimize the risk of ammonia volatilization.
- When the time arrives for the 2nd split ^{15}N fertilizer application, place a plastic sheet/cover around the perimeter of only the second sub-plot of 4 m^2 (this sub-plot will have previously received only ordinary urea) to ensure that ordinary urea is applied only to all areas of the main plot except the 2nd sub-plot during the 2nd fertilizer application. Then, uniformly apply the required amount (0.39 kg) of ordinary urea to the entire 45 m^2 of the larger main plot, but exclude the 2nd sub-plot.
- Remove the plastic sheet, and carefully apply the required amount (34.78 gram) of ^{15}N-labeled fertilizer to the 2nd sub-plot as above.
- Carry out normal farm practices like spraying of herbicides and insecticides, and apply normal irrigation volumes until the wheat crop reaches its maturity.
- At the appropriate time, harvest the wheat crop from each sub-plot. For ^{15}N uptake by below ground (roots) and aboveground plant parts (i.e., stems, leaves and grain), randomly select 3–4 wheat plants from the **middle row** of each sub-plot of ^{15}N; and transfer them to plastic bags. After transporting wheat plant samples to the lab, separate the plant samples into (1) roots, (2) stem and leaves and (3) grain. Wash gently the plant tissue with tap water first, then with distilled water. After washing, allow water to drain and then dry the three types of wheat tissue samples at 65 °C for 7 days or until samples are dried to a constant weight.
- After drying, grind the wheat roots, leaves and stems and grain samples separately to a fine powder (for determination of the total N by Kjeldahl or by the **combustion** method). Then, accomplish the ^{15}N determination by stable isotope mass

spectrometry. Be certain to clean the grinder with a brush (and also use a blower), in between grinding the individual plant tissue samples.

- Also collect four soil samples (each 0–15 cm soil depth) from each of the two sub-plots; mix them to get one composite soil sample for ^{15}N and total N analysis.

Wheat Straw and Grain Yield

- To determine wheat yield, select 3 m × 3 m area within each main-plot (7 m × 7 m) and harvest wheat crop at the same time as above for ^{15}N analysis. Then, separate the biomass into (1) shoot and leaves and (2) grain and record their fresh bulk weight immediately.

[Note: Researchers must not use the small ^{15}N plot for biomass production]

- To determine moisture fraction in leaves plus stems (straw) and in grain, select 2 to 3 randomly chosen wheat plants, from each 3 m × 3 m plot; transfer them to plastic bags, seal each plastic bag using a rubber band to ensure that no water losses occur from the collected plant tissue. After transporting the wheat plant samples to the lab, separate the plant samples into straw and grain, and record their fresh weight. Wash them with tap water to remove the soil. Then take sub-tissue samples of each type of plant tissue (grain and straw), followed by drying the sub-samples of tissue at 65 °C for 7 days.
- Record the dry weights of the plant tissue after 7 days in order to calculate their moisture contents. This will provide the researcher with wheat dry matter yield (**DM**) per hectare as shown in Eq. (6.3).

$$\begin{aligned}&\text{Wheat straw or grain DM (kg per ha)}\\&= \text{FB Wt (kg)} \times \frac{10,000\ \text{m}^2}{\text{harvested area (m}^2)} \times \frac{\text{SD Wt (kg)}}{\text{SF Wt (kg)}}\end{aligned} \tag{6.3}$$

Where, FB Wt is fresh bulk weight (kg per m^2) of the harvested area of the sub-plot (area (3 m × 3 m), and SD Wt and SF Wt are sub-plot sample's dry and fresh weights, respectively.

4.2 *Calculation of Nitrogen Use Efficiency (NUE)*

The following example provides step-by-step guidance for estimating fertilizer 'N use efficiency' of a wheat crop.

A field study was carried out with a wheat crop to assess the fertilizer N use efficiency of wheat grain which received nitrogen fertilizer at the rate of 80 kg N ha $^{-1}$ in 2 split doses (40 kg N ha^{-1} for each of two application times). The experimental sub-plot was 4 m^2 in size and the ^{15}N fertilizer was labeled with exactly 5% atom excess. At the end of growth period, assuming the grain yield from harvested wheat was 2667 kg per ha and the N content in the grain, as obtained by Kjeldahl analysis was 3.0%, the amount of total N removed from the soil by the wheat grain is calculated below (Eq. 6.4):

$$\text{Wheat grain N uptake (kg N per ha)} = \frac{\text{grain yield (kg per ha)} \times \text{total N (\%) of grain}}{100} \tag{6.4}$$

$$\text{wheat grain N uptake} = \frac{2667 \times 3}{100} = 80 \text{ kg N per ha}$$

The grain ^{15}N measurements from the 1st and 2nd split applications of ^{15}N-labeled fertilizers showed that an 'atom excess percentage' of 0.75% and 0.80% occurred, for the two sub-plots. The fertilizer N use efficiency of the grain is calculated as follows:

(i) Percentage grain N derived from 1st and 2nd fertilizer application (*% Ndff*), based on the ratio of grain ^{15}N [0.75% and 0.80%, to fertilizer ^{15}N (5%)], can be calculated from Eq. 6.5.

$$\%Ndff = \frac{^{15}N_{grain}}{^{15}N_{Fertilizer}} \times 100 \tag{6.5}$$

$\%\ Ndff$ for the 1st application $= \frac{0.75}{5} \times 100 = 15\%$
$\%\ Ndff$ for the 2nd application $= \frac{0.80}{5} \times 100 = 16\%$
$\%\ Ndff$ for the two split applications $= 15 + 16 = 31\%$

(ii) From the *% Ndff*, the amount of N derived from the two split fertilizer applications (*Ndff*) is calculated as:

$$Ndff = \%Ndff \times \text{N taken up by crop} \tag{6.6}$$

$$Ndff = \frac{31}{100} \times 80 = 24.8 \text{ kg N per ha}$$

[**Note**: The above equations (Eqs. 6.5 and 6.6) can also be used to calculate *Ndff* of the aboveground wheat plant tissues (straw) as well as roots, if such information is needed.]

Finally, fertilizer N use efficiency *(FNUE)* is calculated from *Ndff* (24.8) and N rate applied (80 kg N ha^{-1}).

$$FNUE = \frac{Ndff}{\text{Total fertilizer } N_{\text{applied}}} \times 100 \tag{6.7}$$

$$FNUE = \frac{24.8}{80} \times 100 = 31\%$$

Thus, in this study the wheat grain derived **31%** of its N from the applied ^{15}N-labeled urea fertilizer, with the remaining N (**69%**) coming from the pre-existing soil N pool.

4.3 An Example for ^{15}N-Labeled Urea Dilution

For diluting 1 kg of ^{15}N-labeled urea with 5 atom–3 atom %, please see the calculations below (Eq. 6.8) using a mixing model based on the following relationship:

$$f_A + f_B = 1 \tag{6.8}$$

Where, f_A and f_B refer to the fractions of labeled fertilizer and un-labeled fertilizers, respectively.

- First calculate the fraction of ^{15}N-labeled fertilizer with 5 atom % (f_A) which will be required for mixing with un-labeled fertilizer to make 3 atom % using Eq. 6.9 below:

$$f_A = \frac{3 - 0.366}{5 - 0.366} = 0.56841 \tag{6.9}$$

- Then calculate the fraction of un-labeled fertilizer using Eq. (6.10) below:

$$f_B = 1 - 0.5684 = 0.43159 \tag{6.10}$$

Thus, for 1 kg of labeled fertilizer with 3 atom %, weigh 0.56841 kg of 5 atom % fertilizer and mix it with 0.43159 kg of un-labeled fertilizer.

5 Biological Nitrogen Fixation (BNF)

Over the past 62 years, world food supplies have become heavily dependent on the use of synthetic N fertilizers predominantly urea, with over half of this N fertilizer being applied to cereal crops. The use of fertilizer N will continue to play a critical role in ensuring world food security. Currently, world fertilizer N use is 113 million metric tons (2016), and this use is expected to increase to 120 million metric tons in 2018. Most of these increases in N fertilizer use will occur in developing countries.

Since the oil crisis of 1974 (and high N fertilizer prices), research attention of many international programs has focused on the use of biological N fixation (BNF) as an alternative N source in agro-ecosystems. Under this natural process, micro-organisms convert atmospheric N (N_2) into ammonia through enzymatic (nitrogenase) reactions for further utilization of the reduced N in plant metabolism. These N_2-fixing micro-organisms can live alone in the soil or in symbiosis with some plant species in a wide range of environments.

A classical example occurring in agricultural systems is the symbiotic association between *Rhizobium* bacteria and the roots of legumes in the Fabaceae family of plants (grain legumes, forage and pasture legumes and a number of tree species). Plant species in the Fabaceae are widely distributed in the world. In this symbiosis, the bacteria inoculate the roots of the legumes, and form nodules which are filled with bacteroids (an altered form of the bacteria).

Legume species are common sources of protein-rich food for humans and feed for their livestock, and they also provide fiber, medicines and other products. Grain legumes can be cultivated in a separate crop rotation, or by intercropping with cereals. The forage legume species are normally used in mixed swards. The tree legume species are employed in agro-forestry and agro-sylvo-pastoral systems. Certain fast growing legume species may be included in cropping systems for use as cover crops, or incorporation into the soil as green manures.

In order to ensure appreciable biological nitrogen fixation (BNF) inputs into agricultural production systems, legume genotypes can be grown from seeds, or propagated vegetatively. Then, selected biofertilizers (commercially available *Rhizobium* cultures) are applied as inoculants to the seeds or seedlings, or to rooted cuttings for tree species. The amount of N_2 fixed by the legumes depends on the symbiosis established between the *Rhizobium* strain and the legume species. Here, the cultivar (genotype) as well as environmental (soil, climate) and agronomic management factors are also important. A number of stress conditions, such as salinity, acidity, drought, extreme temperatures and nutrient deficiencies have negative effects on both partners of the symbiosis.

Appreciable amounts of N_2 are fixed by legumes, thereby contributing to an improved soil fertility status and reducing the need for chemical fertilizer N. A significant proportion of this fixed N is utilized by the cereal crops or grasses which are grown in association with the legumes, or in a crop rotation with the legume. Other apparent benefits called 'legume effects', are also attributed to the inclusion of the legume into the agricultural system. Table 6.2 provides a summary of the legume's effects in agro-ecosystems.

Any program aimed at enhancing the use of legume BNF for improving soil fertility and crop productivity in cropping systems should include the ability to measure N_2 fixation under a wide range of environmental and agronomic management conditions. Methods to assess legume N_2 fixation under field conditions can be grouped into isotopic and non-isotopic methodologies.

5.1 *Estimating Legume BNF Using ^{15}N Isotope Techniques*

Isotopic methods using the stable ^{15}N isotope, both with enrichment and also at natural abundance levels, provide the most sensitive measures of total N_2 fixation over the growing cycle of legume crops. These are also the only methods capable of distinguishing atmospheric N_2 from other sources of N present in the soil.

Of the two main stable isotopes of N, the light isotope ^{14}N, is by far the most abundant (99.6337%). The heavy stable isotope ^{15}N, has an abundance of 0.3663 atom %. If the ^{15}N concentrations within each of the two main sources of N (atmospheric N_2 and soil N) differ appreciably, then it is possible to calculate the

Table 6.2 Main effects of legumes in agro-ecosystem

Issues/processes	Main effect	Details
BNF process *per se*	Soil acidification	Increase in CO_2 fixed/N_2 assimilated
		Soil N uptake also increased
N fertilizer production and application	Reduction in fertilizer N use	Fossil fuel energy use reduced
		CO_2 emissions reduced
		NO_2 emissions reduced
N cycling/N losses	Effects occur during both pre-cropping and cropping	N_2O emissions reduced
Cropping systems		Volatilization as NH_3 reduced
		N leaching reduced
		Usually the NUE of N derived from green manure is lower than N-fertilizer, but large fraction of N-green manure remain in the soil.
	Post-harvest effects	Reduced N_2O emissions,
		NH_3 volatilization, and NO_3^- leaching
		N benefits to next crop/savings from not having to apply as much fertilizer N
	Long-term effects	Soil fertility improvement
		Soil N reserves increased
		Risk of N losses reduced for intensive cropping systems
Use of legume crops	Non-N 'Legume' effects also promoted	Human health improved (quality food diet)
		Biodiversity increased
		Carbon sequestration enhanced
		Soil erosion reduced
		Can interrupt crop pest and disease cycles
		Deep rooting promoted
		Soil structure improved

proportion of the total N that accumulates within the legume tissues that is derived from atmospheric N_2 fixation.

When the aim is the assessment of the N input by N_2-fixing plants through BNF, three parameters are required: the content of N in plant material, the dry matter yield of the N_2-fixing plant and the percentage of N in the N_2-fixing plant derived from the atmosphere (%Ndfa). Considering these three parameters, it is possible to calculate the amount of N fixed, usually expressed in terms of kg N derived from BNF per ha, in field experiments, or mg N derived from BNF per plant or per pot in glasshouse experiments. Based on these estimates, it is also possible to calculate the amount of N derived from soil by discounting the amount of N derived from BNF from the total N.

The %Ndfa depends on the interaction between plant growth and efficiency of microsymbiont strain. It is also depends on the soil physical and chemical properties, (e.g., water and nutrient availability). The two most important isotopic techniques for this purpose are the ^{15}N isotope dilution and ^{15}N natural abundance technique (Boddey et al. 2000; Urquiaga et al. 2012; Collino et al. 2015). Other isotopic

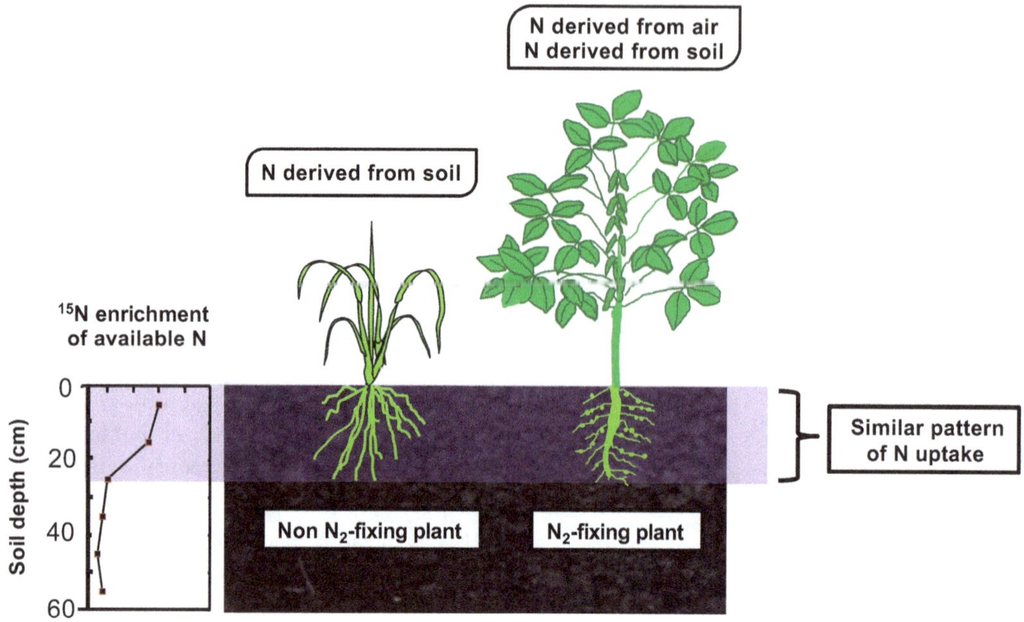

Fig. 6.4 Illustration of the ^{15}N isotope dilution technique for the BNF quantification

techniques, such as $^{15}N_2$ feeding and A-value can be also applied depending on the purpose of the BNF quantification, for which detailed procedures can be found in previous literature (*e.g.*, IAEA 2001).

5.2 ^{15}N Isotope Dilution Technique

The ^{15}N isotope dilution technique has been the most applied isotopic technique for %Ndfa assessment. This technique is based on the dilution of soil N taken up by the N_2-fixing plant by N derived from air through BNF (Fig. 6.4). When this technique is applied it is assumed that the ^{15}N enrichment of non N_2-fixing plant can be used as reference to assess the ^{15}N enrichment of plant-available soil N (Fig. 6.4).

To apply this technique, the soil N taken up by plants is labelled through application of ^{15}N-enriched fertilizers. After the labelling, both N_2-fixing and non N_2 fixing reference plants are grown and sampled at the same time. In fact, if all N forms in soil were easily mineralisable and available for plant uptake, the direct ^{15}N analysis of soil samples could be used as reference to assess ^{15}N abundance of N fraction in N_2-fixing plants derived from soil. However, only the soil mineral N forms (mainly NH_4^+ and NO_3^-), representing a small fraction of N, is available for plant uptake and could theoretically be used to assess the ^{15}N abundance of the N in plants derived from soil (Ledgard et al. 1984; Unkovich et al. 2008). Considering that non N_2-fixing plants has their N nutrition totally dependent on soil mineral N, these plants can be sampled to assess ^{15}N enrichment of the plant-available soil N (Fig. 6.4). In this technique, N_2-fixing plant and the non N_2-fixing plant (reference)

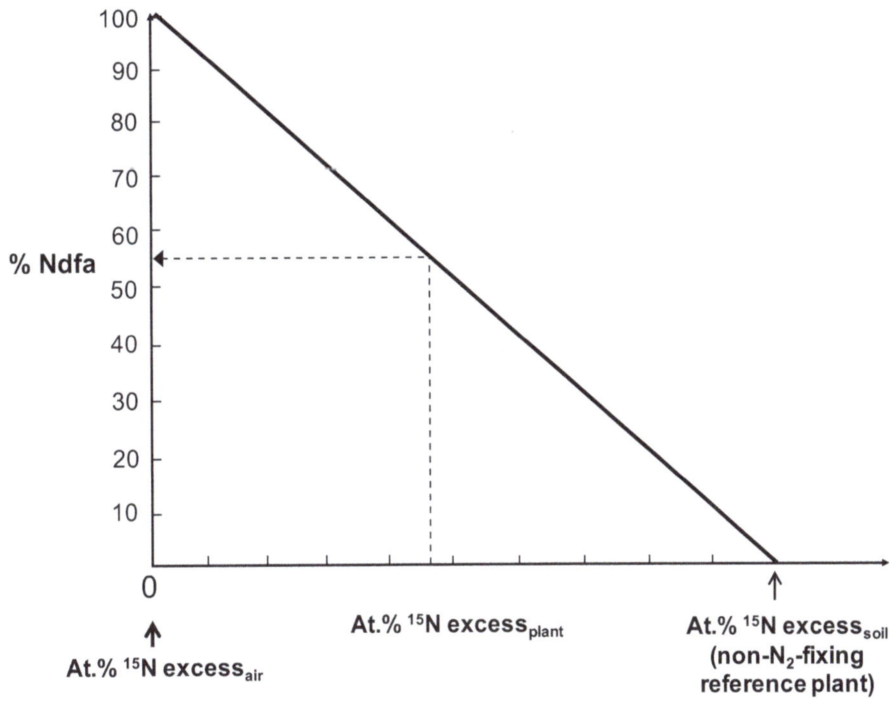

Fig. 6.5 Relationship between ^{15}N enrichment of N_2-fixing plant (abscissa axis) and percentage of N derived from atmosphere (%Ndfa, ordinate axis)

should have similar pattern of N uptake (Fig. 6.4). This is a critical prerequisite for application of ^{15}N isotope dilution technique because, otherwise, the assessment of %Ndfa can be inaccurate when ^{15}N enrichment of soil N is not constant in the time course and/or in the depths of soil N uptake by fixing and non-fixing plants (Baptista et al. 2014; Unkovich et al. 2008). Some procedures can be useful to deal with the non-constant ^{15}N enrichment in time and soil depth, including the use of labile organic materials to immobilise excessive soil mineral N and stabilize N supply over time (Boddey et al. 1995) and constant addition of ^{15}N-labelled fertiliser to the soil (Viera-Vargas et al. 1995). The %Ndfa by N_2-fixing plants is calculated using the following Eq. 6.11:

$$\%\text{Ndfa} = 1 - \frac{\text{atom}\%^{15}\text{N excess}_{\text{N}_2\text{fixing plant}}}{\text{atom}\%^{15}\text{N excess}_{\text{non N}_2-\text{fixing reference plant}}} \times 100 \tag{6.11}$$

The graphical representation of Eq. 6.11 is showed in Fig. 6.5. Taking in consideration that N fertiliser rate can impact the BNF process, it is usual to apply low N rates (e.g., <10 kg N ha^{-1}) when the objective is solely the labelling of plant-available soil N with ^{15}N. When using low rates of N, fertiliser with high ^{15}N enrichment is usually applied to yield plant materials with $^{15}N/^{14}N$ ratios adequate for precise and accurate analyses by spectrometry. The application of 1 kg of ^{15}N excess per hectare (0.1 g ^{15}N excess m^{-2}) usually yields plant materials with sufficient ^{15}N enrichment to be analysed with acceptable precision by most of mass spectrometers (emission spectrometers commonly requires higher ^{15}N enrichments). Considering these values, if a rate of 10 kg N ha^{-1} should be applied, the use of a fertiliser with 10 atom% ^{15}N excess would be recommended. In fact, there is a possibility of using lower ^{15}N enrichments depending on the spectrometer type, but this must be based on a rigorous assessment of analytical precision and after significant experience was gained. When this methodology is used for woody perennials, higher N rates (e.g., 20 kg N ha^{-1}) and/or ^{15}N enrichments should be used.

The selection of non N_2-fixing plants is a very important step for the accurate quantification of BNF by ^{15}N isotope dilution technique. Some recommendations are presented below to avoid some biases due the selection of non N_2-fixing reference plants:

- To be sure that the reference plants do not have the ability of N_2-fixing, which could be identified by:

  Classical N deficiency symptoms (*e.g.*, pale green or yellow colour, especially in the older leaves).

  Literature search indicating the inability of N_2-fixing. That is especially important for Poaceae, considering that some species of this plant family has the ability of N_2-fixing (Urquiaga et al. 1992; Reis et al. 2001).

  Absence of nodules when non-nodulating isolines or non-inoculated legumes are used as reference plants.

- To use three or more reference plant species to assess the variability associated with the ^{15}N enrichment of plant available soil N.
- Select non reference plants that presents patterns of N uptake similar to that of N_2-fixing plant, that is, have similar rooting depth and architecture exploiting the same pool of plant-available soil N and have the same dynamics of N uptake over time;
- If different varieties of a N_2-fixing crop having significant different life cycles are to be compared for the BNF ability, the group of varieties with similar life cycle must be paired with a reference plants with the duration of growth.
- Considering that differences in soil history can affect N mineralisation dynamics, additional reference plants must be grown and sampled for each different crop sequence even when BNF will be assessed for only one N_2-fixing crop type (e.g., effect of cropping history on BNF associated to soybean);
- Ideally, each reference plant should be considered as an additional treatment in the layout of field and glasshouse experiments, that is, they should be grown in additional field plots with the same replication and randomisation made for N_2-fixing crops.

To apply ^{15}N-fertilisers aiming to label the plant-available soil N, the same strategy of ^{15}N-microplot inside the main field plot previously described to study Fertilizer Use Efficiency can be used for BNF quantification using ^{15}N isotope dilution technique. The plant material sampled in micro-plots will provide an estimate of %Ndfa. The dry matter yield, the total N taken up and the amount of N derived from BNF (e.g., kg N-BNF ha^{-1}) can be measured by harvesting larger area of the plot, including the area that received ^{14}N-fertiliser.

5.3 *Calculation of the Amount of N Derived from BNF by ^{15}N Isotope Dilution Technique*

The following example shows the steps for estimating the %Ndfa the amount of N derived from BNF, in kg N ha^{-1}, for soybean crop by ^{15}N isotope dilution technique. A field study was carried out with a commercial soybean variety to assess the performance of three *Rhizobium* strains under a condition of water stress. The soybean was sown at row spacing of 0.50 m and three plant species were included as non N_2-fixing reference plants: *Sorghum* sp.; *Brassica* sp. and non-nodulating soybean. The quantification of BNF will be performed by ^{15}N isotope dilution technique. Each experimental plot was 36 m^2 (6 m $\times$ 6 m). A micro-plot was established in an area of 9.0 m^2 (3.0 m $\times$ 3.0 m) in each experimental plot. For this study, ^{15}N-labeled ammonium sulphate ($(^{15}NH_4)_2SO_4$) with enrichments of 20 atom % ^{15}N in excess was applied 50 days before sowing to each micro-plot at a rate of 5 kg N ha^{-1}. Non-labelled fertiliser ($(^{14}NH_4)_2SO_4$) was also applied the remaining area of the plot. The soybean and reference plants were sown and harvested (105 days after sowing) concomitantly. The plants (shoot tissue) corresponding 1.5 m of the central row of the ^{15}N-labelled micro-plot were collected, weighted, oven-dried, reweighted, ground and analysed for total N and ^{15}N. Dry mass, N content and ^{15}N-erichment are presented below (Table 6.3).

The mean value of ^{15}N enrichment of reference plants was 1.1305 atom % ^{15}N excess. An example calculation for the soybean inoculated with strain A is presented as follows using Eqs. 6.12 and 6.13:

$$\text{Total N in shoot (kg per ha)} = \frac{\text{Dry mass (kg per ha)} \times \text{N content (\%)}}{100} \tag{6.12}$$

Table 6.3 Example of results of field experiment with soybean for quantification of BNF by ^{15}N isotope dilution technique

Parameter	Soybean inoculated with strain A	Soybean inoculated with strain B	Soybean inoculated with strain C
Dry mass (kg ha^{-1})	5097	4850	3105
N content (%)	3.7	3.9	3.7
Atom % ^{15}N excess	0.1420	0.0330	0.0920

$$\text{Total N in shoot} = \frac{5097 \times 3.7}{100} = 189 \text{ kg N per ha}$$

$$\%\text{Ndfa} = 1 - \frac{\text{atom}\%^{15}\text{N excess}_{\text{N}_2\text{fixing plant}}}{\text{atom}\%^{15}\text{N excess}_{\text{non N}_2-\text{fixing reference plant}}} \times 100$$

$$\%\text{Ndfa} = 1 - \frac{0.1420}{1.1305} \times 100 = 87\%$$

$$\text{Amount of N derived from BNF (kg per ha)} = \frac{\text{Total N in shoot (kg per ha)} \times \%\text{Ndfa}}{100} \quad (6.13)$$

$$\text{Amount of N derived from BNF} = \frac{189 \times 87}{100} = 165 \text{ kg N per ha}$$

Considering the other data of shoot dry mass, N content and atom % ^{15}N excess, the amounts of N derived from BNF for soybean inoculated with strain B was 184 kg N ha^{-1} and for soybean inoculated with strain C was 106 kg N ha^{-1}.

5.4 *^{15}N Natural Abundance Technique*

This technique depends on the slight natural enrichment of ^{15}N in the soil, relative to atmospheric N_2. The slight increase of ^{15}N in soil is a consequence of the non-identical behaviour of the light and heavy isotopes involved in various reactions in the soil environment. The ^{15}N isotopic fractionation, also called the mass discriminatory effect (Xing et al. 1997), is a result of complex and prolonged interaction of biological, chemical and physical processes in soils, which results in fractionation between ^{15}N and ^{14}N. There is a tendency of the reaction products, such as the gaseous N forms produced by denitrification, to become relatively enriched in the lighter isotope ^{14}N, while the remaining N compounds, which can be stabilised in soil organic matter over time, tend to be enriched in the heavier isotope ^{15}N (Xing et al. 1997). It is important to consider that this small ^{15}N enrichment occurs in a long time scale, and is closely associated to soil organic matter retention and long-term dynamics (Ledgard et al. 1984).

Considering that ^{15}N natural abundance technique is based on the analyses of plant samples having very small ^{15}N deviation relative to atmospheric N_2, it is usual to express the results of ^{15}N natural abundance analyses in terms of δ units. The δ^{15}N value is the difference in the ratio ^{15}N:^{14}N of a given sample and the ratio ^{15}N:^{14}N in the nominated international standard of atmospheric N_2, expressed by parts per thousand (‰). One unit of δ^{15}N (1.0‰) is a thousandth of the ^{15}N natural abundance of the atmosphere (0.3663 atom% ^{15}N) above or below the natural abundance of atmospheric N_2, that is, one unit of δ^{15}N it is equal to 0.0003663 atom% ^{15}N excess. The following Eq. 6.14 is applied to calculate the δ^{15}N:

$$\delta^{15}N(‰) = \frac{atom\%^{15}N_{sample} - atom\%^{15}N_{atmosphere}}{atom\%^{15}N_{atmosphere}} \times 1000 \quad (6.14)$$

Therefore, the $\delta^{15}N$ of atmospheric N_2 will be by definition equal to 0‰. Positive value of $\delta^{15}N$ means that there are an enrichment of ^{15}N in the sample compared to the atmospheric N_2 and negative values means that the sample presents a slightly depletion. For example, if a plant sample has 0.35855 atom% ^{15}N, the resulting $\delta^{15}N$ of this sample is:

$$\delta^{15}N = \frac{0.3659 - 0.3663}{0.3663} \times 1000 = -1.09‰$$

The main advantage of the natural abundance technique, compared to ^{15}N isotope dilution technique, is the no requirement to add ^{15}N fertiliser to label the soil available N, which is a very expensive consumable and, depending on the N rates, it can affect BNF process. However, an important disadvantage of this technique is the need for an Isotope Ratio Mass Spectrometer with high precision.

The Eq. 6.15 can be used to calculate the Ndfa% by using the ^{15}N natural abundance technique is:

$$Ndfa\% = \frac{\delta\,^{15}N_{reference\ plant} - \delta\,^{15}N_{fixing\ plant}}{\delta\,^{15}N_{reference\ plant} - B} \times 100 \quad (6.15)$$

where B is the $\delta^{15}N$ for the N_2 fixing plant when completely dependent on N_2 fixation for growth. The B value is usually negative as a result of isotopic fractionation within the legume. The value of B depends on the plant species, plant age, symbiont and growth conditions. Unkovich et al. (2008) presented some tables with compilation of a wide number of B values for shoot of many tropical and temperate legumes, which can be used to estimate %Ndfa with an acceptable accuracy depending on the N_2 fixation level.

Another important factor affecting the %Ndfa estimate is the $\delta^{15}N$ of the reference plant. The higher is this parameter, the better is the estimate of %Ndfa because this will result in less impact of biases associated to small variability of some processes, such as the mineralisation intensity of soil N pools, isotopic discrimination in plants or small differences in root architecture between N_2-fixing and non-fixing reference plants. Reference $\delta^{15}N$ higher than 4‰ have been considered suitable for estimating %Ndfa in N_2-fixing plants (Unkovich et al. 2008).

An important practical procedure to have an initial estimate of ^{15}N natural abundance of plant-available soil N before the beginning of the experiment is the ^{15}N analysis of non N_2-fixing broadleaf and grass weeds in the experimental area available for BNF studies. Separated samples of the different reference plant should be collected in different points of the area to assess the variability of $\delta^{15}N$ in plant-available N (not a composite sample). In addition to that, details on the history of the area are very useful, including previous crop type, N fertilisation (type and rates) and use of inoculants.

Table 6.4 Example of results of glasshouse experiment for measuring BNF associated to *Phaseolus vulgaris* by ^{15}N natural abundance technique

Parameter	Common bean variety A	Common bean variety B
Dry mass (g per pot)	45	39
N content (%)	2.5	2.6
$\delta^{15}N$	0.52	0.96

All recommendation presented for ^{15}N isotope dilution technique to select non N_2-fixing plants must also be considered for the ^{15}N natural abundance technique. The reference plants must be considered as additional treatments in the experimental design, with replication and randomisation. When experiments are conducted as randomised block design the %Ndfa estimate for the plants of a given block should be performed with the $\delta^{15}N$ of the references of the same block individually.

Calculation of the Amount of N Derived from BNF by ^{15}N Natural Abundance Technique The following example shows the steps for estimating the %Ndfa and the amount of N derived from BNF, in kg N ha^{-1}, for common bean (*Phaseolus vulgaris, L.*) by ^{15}N natural abundance technique:

A glasshouse study was carried out with two varieties of common bean to assess the osmotic effect of a salt (NaCl) on the BNF performance. The common bean cultivars and three reference plants (*Sorghum* sp.; *Brassica* sp. and non-nodulating bean) were sown in 10-L pots with 10 kg of soil. Three plants were used per each pot. Soil salinity was simulated by adding NaCl solution in soil. The BNF quantification will be performed by ^{15}N natural abundance technique. The common bean shoots were collected at 60 days after sowing, weighted, oven-dried, reweighted, ground and analysed for total N and ^{15}N. Dry mass, N content and ^{15}N abundance are presented below (Table 6.4).

The mean value of $\delta^{15}N$ of reference plants was 9.82‰ and the B value used for common bean was −1.97‰. An example calculation for the variety A is presented as follows:

$$\text{Total N in shoot (g per pot)} = \frac{\text{Dry mass (g per pot)} \times \text{N content (\%)}}{100}$$

$$\text{Total N in shoot} = \frac{45 \times 2.5}{100} = 1.13 \text{ g N per pot}$$

$$\text{Ndfa\%} = \frac{\delta\,^{15}N_{\text{reference plant}} - \delta\,^{15}N_{\text{fixing plant}}}{\delta\,^{15}N_{\text{reference plant}} - B} \times 100$$

$$\text{Ndfa\%} = \frac{9.82 - 0.52}{9.82 - (-1.97)} \times 100 = 79\%$$

$$\text{Amount of N derived from BNF (mg per pot)} = \frac{\text{Total N in shoot (g per pot)} \times \%\text{Ndfa}}{100}$$

$$\text{Amount of N derived from BNF} = \frac{1.13 \times 79}{100} \times 1000 = 893 \text{ mg N per pot}$$

Considering the other data of shoot dry mass, N content and atom % ^{15}N excess, the amounts of N derived from BNF for variety B was 761 mg N per pot.

5.5 *Correction for N Derived from Seed*

In some experiments using plants with proportionally large seeds or when plants are sampled in early growth stages, when N derived from seeds can supply a significant proportion of plant N, a correction in ^{15}N enrichment/abundance of plant materials can improves the accuracy of the %Ndfa estimate (Okito et al. 2004). This correction is made by subtracting the amounts of N derived from seed and its ^{15}N enrichment/ abundance from plant material. For example, the following Eq. 6.16 is applied for correction when ^{15}N natural abundance is applied:

$$\delta^{15}N_{\text{plant (SC)}} = \frac{(\%N_{\text{plant}} \times DM_{\text{plant}} \times \delta^{15}N_{\text{plant}}) - (\%N_{\text{seed}} \times DM_{\text{seed}} \times P_s \times \delta^{15}N_{\text{seed}})}{(\%N_{\text{plant}} \times DM_{\text{plant}}) - (\%N_{\text{seed}} \times DM_{\text{seed}})} \tag{6.16}$$

where SC indicates the correction for seed N, %N is the N content, DM is the dry mass, P_s is the proportion of the seed N assimilated by plant tissue. P_s is usually assumed to be 0.5 when shoot tissue is analysed considering that half of N seed is incorporated in into the aerial tissue. The same equation can be applied for ^{15}N isotope dilution technique by replacing the values of $\delta^{15}N$ by atom% $\delta^{15}N$ excess. When plants grown under field conditions and are sampled at the maturity stage this correction does not usually have a significant influence in the final estimate of % Ndfa because the contribution of seed N in this case is commonly small.

General Comments:
The use of ^{15}N techniques has been successfully applied to measure BNF in many agricultural systems in many regions of the world. However, before the beginning of the experimentation using those isotope techniques it is important to take into account the main requirements needed for success in the BNF measurement: (i) the requirement of highly skilled workers for all activities from the selection of the experimental area to the interpretation of the ^{15}N analysis, and (ii) the requirement of financial resources considering that consumables for ^{15}N analysis are usually expensive compared to other routine plant and soil analyses. The selection of the most appropriate technique will depend mainly on the precision of the Mass-

Table 6.5 Some advantages (A) and disadvantages (D) of two ^{15}N isotope techniques for measuring BNF in agricultural systems

Criteria	^{15}N isotope dilution technique	^{15}N natural abundance technique
Requirement of reference plants	D	D
Cost with ^{15}N fertiliser	D	A
Cost with ^{15}N analysis	D	D
Requirement of high-skilled technicians	D	D
Requirement of high-precision spectrometers	A	D
Application of the technique in areas (with grown plants) not initially designed for BNF assessment (*e.g.*, farms, natural systems)	D	A
Need of considering isotope fractionation (B value)	A	D
Field variability of soil ^{15}N	A	D
Application in perennial systems	A	A
Application in experiments with soils presenting plant-available N with low $\delta^{15}N$ (<4‰)	A	D
Time integrated measurement of %Ndfa	A	A
Measurement of amount of N derived from BNF per area (field) or per pot (glasshouse)	A	A

Spectrometer used for ^{15}N analysis of plant materials. Other criteria are also presented in Table 6.5.

6 Water Stable Isotope Technique to Determine Evapotranspiration Partitioning

In agriculture, evapotranspiration (ET), or the flux of water from a vegetated surface via both evaporation (E) and transpiration (T) by plants, is an important component of the water budget. Water loss via transpiration can be considered 'good' water use, while water loss via evaporation can be considered 'wasted' water use (Fig. 6.6). Transpiration occurs through stomatal pores, the pores which are also used by the plants for uptake of atmospheric CO_2 in photosynthesis, and subsequent biosynthesis of carbon compounds, a process which ultimately leads to biomass gain. Stomata are tightly controlled by plant physiological signals to optimize carbon gain per unit of water lost. The use of the stable isotopes ^{18}O and ^{2}H as signatures in water and water vapor can help scientists to differentiate between water losses through direct soil evaporation versus transpiration from the plant leaves. That knowledge can be used to apply appropriate soil and water conservation strategies such as minimum tillage, mulching and a drip/spray irrigation system in order to minimize soil evaporation under a range of different management practices. Water use efficiency

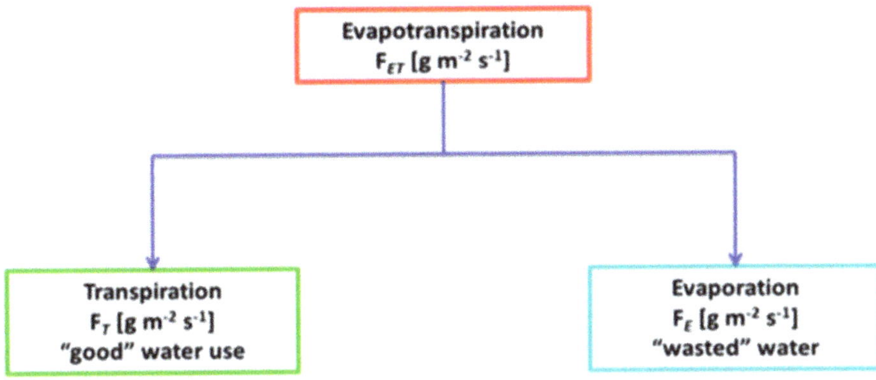

Fig. 6.6 Evapotranspiration model

(WUE) of a plant species or crop type is related both to the plant's genetics, as well as acclimation by the plant to the irrigation regime.

Historically, the characterization of the plant processes involved in transpiration was performed through cumbersome and inaccurate water flux measurements. However, with the recent advancement of laser-based water vapor isotope analyzers, various calculation models have been developed to correlate the real-time, spatial, and temporal isotopic measurements with evaporation and transpiration fluxes (F_{ET} and F_T).

According to Yakir and Sternberg (2000), the ratio of these fluxes is calculated using Eq. 6.17:

$$f_{T/ET} = \frac{F_T}{F_{ET}} = \frac{\delta_{ET} - \delta_E}{\delta_T - \delta_E} \tag{6.17}$$

Where, δ_{ET} is the isotopic composition of bulk evapotranspiration, δ_E is the isotopic composition of evaporated soil-water, and δ_T is the isotopic composition of water transpired by the plant.

In this section, we demonstrate how laser-based absorption spectroscopy, and in particular, Cavity Ring-Down Spectroscopy (CRDS), can be applied to many steps of ET analyses, including: (i) characterization of partial pressure and the isotopic composition of the vertical water vapor profiles to determine the bulk ET signal through a Keeling mixing model, (ii) the use of soil water isotopic composition, in combination with the Craig-Gordon model, to determine the evaporation flux signature, and (iii) direct measurement of the isotopic signature of transpiration occurring in leaf chambers in order to determine the isotope signature of the water source.

6.1 Determining δ_{ET} Using the Keeling Mixing Model

6.1.1 Theory

The isotopic composition of an evapotranspiration flux can be determined by using the Keeling mixing model (1958), a model which correlates water concentration

(C) and the isotopic composition (δ) of the mixed air above the surface (A), the background air (B), and the evapotranspiration flux (ET) Eq. 6.18.

Keeling Mixing Model

$$C_A.\delta_A = C_B.\delta_B + C_{ET}.\delta_{ET} \tag{6.18}$$

Assuming the concentration and the isotopic composition of the background air (C_B, δ_B) and evapotranspiration (C_{ET}, δ_{ET}) are constant over a short period of time, Eq. (6.18) can be rearranged so that δ_A is a function of $1/C_A$. In this case, (Eq. 6.19) the intercept of a plot of $1/C_A$ (x-axis) versus δ_A (y-axis) will yield δ_{ET}.

$$\delta_A = (\delta_B - \delta_{ET})\frac{C_B}{C_A} + \delta_{ET} \tag{6.19}$$

6.1.2 Experimental Approach

Experimentally, one can measure the isotopic composition of the mixed air, δ_A, at various concentrations, C_A, by sampling the air at different elevations above the surface. The vertical profile provides the water concentration gradient which is required in order to determine δ_{ET}.

The procedure involves the following steps.

- Sample air above the soil surface at different heights. The heights at which you sample will depend on the specifics of the ecosystem being studied.
- Connect the sample lines to a manifold using a rotary valve selector.
- If possible, use a rotary valve which can be controlled via a Picarro water isotope analyzer. For example, the Picarro L2130-*i* or L2140-*i*, can be used to select the sample line through which air will be sent to the analyzer.
- Run the analyzer in dual mode: vapor and liquid measurement allows the analyzer to self-calibrate using a liquid water standard, while the vapor mode analyzes the sampled water vapor, thereby providing isotopic composition and concentration.
- Using the analyzer's Dual Mode Coordinator, set the system to measure vapor from each sample port for 10 min (i.e., a total of 50 min for one cycle – please note 5 sampling heights in this example, Fig. 6.7). Measurements should be made at a frequency of 1 Hz.
- It is recommended that the analyzer be calibrated with liquid water standards of a known isotopic composition once every 8 h. The auto-sampler injects the liquid standard sample into the vaporizer. Each injection measurement takes 9 min and a minimum of 6 injections for each liquid standard is required.

After the analyzer measurement, results are collected and processed (averaging and normalizing for each calibration), δ_A and $1/C_A$ are plotted on a graph as shown in the Fig. 6.8. Note that δ_{ET} is the y-intercept of the regression line between δ_A and $1/C_A$.

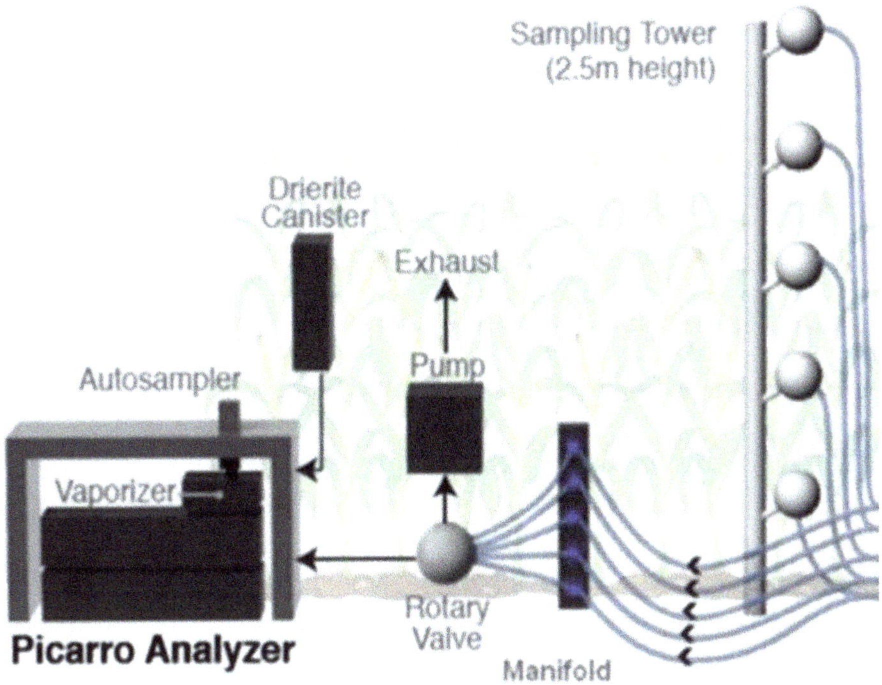

Fig. 6.7 Example of experimental setup for sampling water vapor at different heights

Fig. 6.8 Example of a Keeling plot derived from a vertical profile of 5 water vapor measurements

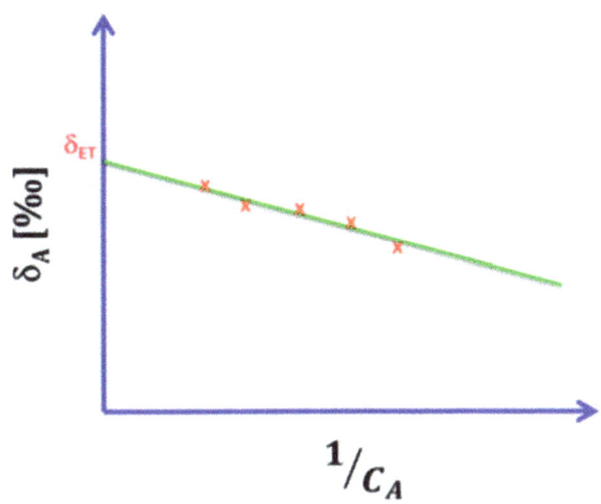

6.2 *Determining δ_{ET} Using the Craig-Gordon Model*

6.2.1 Theory

The Craig-Gordon model (1965) is used to estimate the isotopic composition of soil-water evaporation. The model takes into account the effect of equilibrium and kinetic fractionations during the phase change between liquid to vapor (Eq. 6.20).

$$\delta_E = \frac{(\delta_L \alpha_e h_s - h'_A \delta_A) - (h_s - h_s \alpha_e) - (\varepsilon_k)}{(h_s - h'_A) + \varepsilon_k} \tag{6.20}$$

Where, α_e is the equilibrium vapor-liquid fractionation factor. It can be calculated as a function of soil temperature, T_s, [K] as explained by Majoube (1971) (Eq. 6.21).

For ^2H

$$\ln \alpha_e = -52.612.10^{-3} + \frac{76.248}{T_s} - \frac{24.844.10^3}{T_s^2} \tag{6.21}$$

For ^{18}O (Eq. 6.22)

$$\ln \alpha_e = 2.0667.10^{-3} + \frac{0.4156}{T_s} - \frac{1.137.10^3}{T_s^2} \tag{6.22}$$

Where,

- δ_L is the soil liquid water isotopic composition [‰]
- δ_A is the ambient air water vapor isotopic composition [‰]
- h_s is the soil vapor saturation which is defined by Mathieu and Bariac (1996) (Eq. 6.23):

$$h_s = e^{M\varphi_s/RT_s} \tag{6.23}$$

- M is the molecular weight of water (18.0148 g/mol)
- φ_s is the soil potential (matric potential) of the evaporating surface [kPa]
- R is the ideal gas constant (8.3145 mL MPa/mol/K)
- T_s is the soil temperature, i.e. the temperature of the evaporating surface [K]
- ε_k is the kinetic isotopic fractionation factor (Eq. 6.24)

$$\varepsilon_k = n(h_s - h'_A)\left(1 - \frac{D_i}{D}\right) \tag{6.24}$$

- D_i/D, the ratio of molecular diffusion coefficients of water vapor in dry air, is taken as 0.9757 from Merlivat (1978) (Eq. 6.25):

$$h'_A = \frac{h_A e_{sA}}{a_w e_{s0}} \tag{6.25}$$

- h'_A is the humidity of the atmosphere normalized to the evaporating surface
- h_A is the humidity of the atmosphere
- e_{sA} and e_{sO} are the saturation vapor pressures at the atmosphere's (air's) temperature and the temperature of the evaporation surface, respectively
- a_w is the thermodynamic activity of water

- n is related to the volumetric soil moisture (θ_s), the moisture of the residual (θ_{res}) and the saturated moisture (θ_{sat}), as proposed by Mathieu and Bariac (1996) (Eq. 6.26):

$$n = 1 - \frac{1}{2}\left(\frac{\theta_S - \theta_{res}}{\theta_{sat} - \theta_{res}}\right) \tag{6.26}$$

6.2.2 Experimental Approach

Measuring δ_L

The isotopic composition of soil water will be measured using a Picarro water isotope analyzer (Fig. 6.9).

Several water extraction methods are available, as below:

Cryogenic Distillation

Cryogenic distillation is an established technique for extracting liquid water from samples, for example soils and leaves. Once extracted, the liquid water can be analyzed for its isotopic composition using a High Precision Vaporizer and Picarro water isotope analyzer.

Picarro Induction Module (IM)

The Picarro IM extracts water from soil samples by inductively heating the sample and directly sending the evaporated water vapor to the Cavity Ring-Down Spectroscopy (CRDS) analyzer. Prior to analysis on the CRDS, the water vapor is passed through a micro combustion cartridge to remove organic molecules which could potentially interfere with CRDS analysis. For more information about Picarro's Induction Module, please visit:

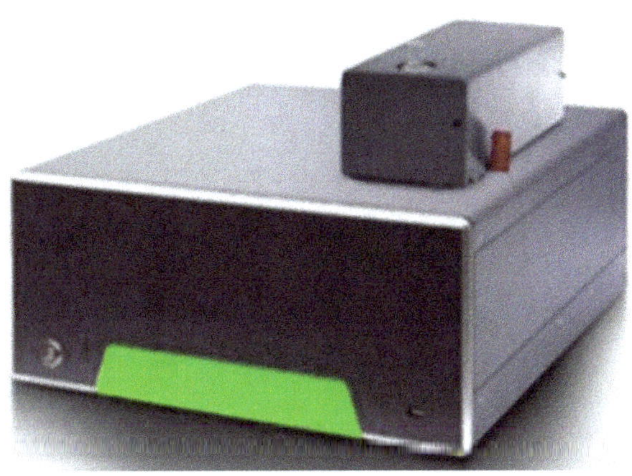

Fig. 6.9 Induction Module and Isotopic Water Analyzer

http://www.picarro.com/isotope_analyzers/im_crds.

When extracting water from soils using either of the above methods, caution should be applied to ensure that water extraction is complete. If water extraction is not complete, it is possible that fractionation may occur during isotopic analysis (or during the extraction process), which could then lead to inaccurate results. Care should also be taken during the storage of soil samples.

Measuring δ_A and C_A and Determining h_A

The isotopic composition of water vapor in the background ambient air is measured with the CRDS water analyzer when it is in the vapor mode.

Sample the ambient air well away from the studied system to ensure that no 'local' water vapor contamination occurs from evapotranspiration of the experimental plot, thereby affecting the ambient air measurement. This can be accomplished by placing the CRDS analyzer input port at an appreciable distance away from the experimental plot, or by connecting tubing to the inlet port of the CRDS in order to collect the air from well-above the canopy. The specific height above the canopy will be dependent on the ecosystem being studied.

Ensure that the CRDS analyzer is calibrated for isotopic composition and also the concentration dependence of the isotopic composition. For information on how to calibrate a Picarro Water Isotope Analyzer refer to the User's Manual. A recent version is available at:

https://picarro.box.com/s/0nh2wvm4n4ojf8jlmj7v.

The CRDS analyzer should be operated in dual mode: vapor and liquid. The liquid measurement allows the analyzer to calibrate itself with liquid standards. The vapor mode analyzes the sampled water vapor in ambient air to provide isotopic composition δ_A and concentration C_A.

Calculate h_A using C_A.

6.3 *Determining δ_T via Direct Measurement at the Leaf*

6.3.1 Theory

When re-arranging the mass balance established in Eq. 6.18, we get (Wang et al. 2012):

$$\delta_T = \frac{C_M\delta_M - C_A\delta_A}{C_M - C_A} \tag{6.27}$$

Where, δ_A and C_A are the isotopic composition and water concentration of the ambient air; δ_M and C_M are the isotopic composition and water concentration measured from the leaf chamber, i.e. where transpiration water vapor mixes with ambient air.

Fig. 6.10 Experimental setup for measuring δ_M and C_M

6.3.2 Experimental Approach

Measuring δ_A and C_A

Follow exactly the procedure described on the previous page. One can directly measure the isotopic composition of the mixed air, δ_M and water concentration, $C_{M,}$ inside of the leaf chamber. Figure 6.10 depicts the experimental setup:

- A leaf chamber is typically made of transparent plastic with a variable internal volume which will be dependent on the leaf size. The chamber has two small air vents to allow ambient air to flow into the chamber and mix with the water vapor generated by transpiration from the leaf.
- A 1/8-inch ID Teflon tubing connects the leaf chamber to the analyzer.
- Place a leaf, which remains attached to the plant, into the leaf chamber.
- Ensure that the CRDS analyzer is calibrated for isotopic composition and concentration dependence of the isotopic composition. For information on how to calibrate the Picarro Water Isotope Analyzer refer to the User's Manual. A recent version is available at:
  https://picarro.box.com/s/0nh2wvm4n4ojt8jlmj7v.
- As detailed previously, operate the analyzer in the dual measurement mode: liquid and vapor. The liquid measurement allows the analyzer to calibrate itself with a liquid water standard while the vapor mode analyzes the sampled water vapor to provide isotopic composition and concentration.

7 Application of Other Isotopes

As mentioned in earlier sections of this chapter, nuclear and isotopic techniques have a wide range of applications in the soil-water-plant interaction studies, covering the fields such as plant ecology, physiology, biochemistry, nutrition, microbiology, protection against insect pests, and soil fertility, chemistry, physics, and hydrology, etc. Few common examples of the applications of isotopic and nuclear techniques in agricultural research are listed below.

- ^{32}P fertilizer use efficiency, root activity, DNA probes in molecular biology
- ^{35}S in soil and fertilizer studies
- ^{65}Zn in plant uptake and use efficiency
- ^{13}C, ^{14}C in soil organic matter dynamics, root activity, photosynthesis, pesticide residues, water use efficiency, etc.
- ^{22}Na, ^{36}Cl, ^{40}K in ion uptake and mechanism of salt tolerance in plants
- ^{137}Cs in soil erosion studies
- ^{60}Co for sterile insects in integrated pest management (IPM)
- ^{198}Gold-198 for detection of termite colonies in agricultural fields

The nuclear and isotopic techniques are the supporting tools, and not substitute, to the conventional techniques for understanding the biological processes and mechanisms of ecosystem functioning. Therefore, a careful evaluation is required with regard to: i) the need for using an isotopic/nuclear technique, and ii) the choice of the appropriate isotopic/nuclear considering the research objective, facilities and expertise available, risks involved in safe handling and disposal of hazardous materials, and the financial considerations. In this context, the stable isotopes are the ever preferred choice in soil-water-plant-atmosphere studies. Thus, examples and protocols of using ^{15}N, ^{18}O and ^{2}H in plant nutrient and water use efficiency studies have been elaborated in this chapter. The reader is, however, referred to the IAEA Training Manuals (IAEA 1990, 2001) and the review by Nguyen et al. (2011), may one need further details.

References

Baptista RB, Moraes RF, Leite JM, Schultz N, Alves BJR, Boddey RM, Urquiaga S (2014) Variations in the ^{15}N natural abundance of plant-available N with soil depth: their influence on estimates of contributions of biological N_2 fixation to sugarcane. Appl Soil Ecol 73:124–129

Boddey RM, Oliveira OC, Alves BJR, Urquiaga S (1995) Field application of the ^{15}N isotope dilution technique for the reliable quantification of plant-associated biological nitrogen fixation. Fertil Res 42:77–87

Boddey RM, Peoples MB, Palmer B, Dart PJ (2000) Use of the ^{15}N natural abundance technique to quantify biological nitrogen fixation by woody perennials. Nutr Cycl Agroecosyst 57 (3):235–270

Collino DJ, Salvagiotti F, Perticari A, Piccinetti C, Ovando G, Urquiaga S, Racca RW (2015) Biological nitrogen fixation in soybean in Argentina: relationships with crop, soil, and meteorological factors. Plant Soil 392(1–2):239–252

Craig H, Gordon L (1965) Deuterium and oxygen-18 variations in the ocean and the marine atmosphere. In: Stable isotopopes in oceanographic studies and paleotemperatures. Laboratorio Di Geologica Nucleare, Pisa, pp 9–130

IAEA (1990) Use of nuclear techniques in studies of soil-plant relationships. Training course series no 2 (Hardarson G, ed). International Atomic Energy Agency, Vienna, Austria, 223 pp

IAEA (2001) Use of isotope and radiation methods in soil and water management and crop nutrition – Training Course Series No. 14. International Atomic Energy Agency, Vienna, Austria, 247 pp

Keeling C (1958) The concentration and isotopic abundances of atmospheric carbon dioxide in rurals. Geochimica et Cosmochimica Acta 13(4):322–324

Ledgard SF, Freney JR, Simpson JR (1984) Variations in natural enrichment of 15N in the profiles of some Australian pasture soils. Soil Res 22:155–164

Majoube M (1971) Fractionnement en oxygene-18 et en deuterium entre l'eau et sa vapeur. J Chim Phys Biol 68:1423–1436

Mathieu R, Bariac T (1996) A numerical model for the simulation of stable isotope profiles in drying soils. J Geophys Res 101(D7):12685–12696

Merlivat L (1978) Molecular diffusivities of $H_2{}^{16}O$, $HD^{16}O$, and $H_2{}^{18}O$ in gases. J Chem Phys 69:2864–2871

Nguyen ML, Zapata F, Lal R, Dercon G (2011) Role of isotopic and nuclear techniques in sustainable land management: achieving food security and mitigating impacts of climate change. In: Lal R, Stewart BA (eds) World soil resources and food security, advances in soil science, vol 18. CRC Press, Boca Raton, pp 345–418

Okito A, Alves BRJ, Urquiaga S, Boddey RM (2004) Isotopic fractionation during N_2 fixation by four tropical legumes. Soil Biol Biochem 36(7):1179–1190

Peoples MB, Faizah AW, Rerkasem B, Herridge DF (1989) Methods of evaluating nitrogen fixation by nodulated legumes in the field, ACIAR Monograpg No 11. ACIAR, Canberra, 76 pp

Pessarakli M (1991) Dry matter nitrogen-15 absorption and water uptake by green beans under sodium chloride stress. Crop Science 31:1633–1640

Reis VM, Reis FB Jr, Quesada DM, Oliveira OC, BJR A, Urquiaga S, Boddey RM (2001) Biological nitrogen fixation associated with tropical pasture grasses. Funct Plant Biol 28 (9):837–844

Saranga Y, Menz M, Jiang CX, Robert WJ, Yakir D, Andrew HP (2001) Genomic dissection of genotype X environment interactions conferring adaptation of cotton to arid conditions. Genome Res 11:1988–1995

Unkovich M, Herridge D, Peoples M, Cadisch G, Boddey RM, Giller K, Alves BJR, Chalk P (2008) Measuring plant-associated nitrogen fixation in agricultural systems, ACIAR Monograph no 136. Australian Centre for International Agricultural Research, Canberra, p 258

Unkovich MJ, Pate JS (2000) An appraisal of recent field measurements of symbiotic N_2 fixation by annual legumes. Field Crops Res 65(2):211–228

Urquiaga S, Cruz KHS, Boddey RM (1992) Contribution of nitrogen fixation to sugar cane: nitrogen-15 and nitrogen balance estimates. Soil Sci Soc Am J 56:105–114

Urquiaga S, Xavier RP, Morais RF, Batista RB, Schultz N, Leite JM, Sá JME, Barbosa KP, Resende AS, Alves BJR, Boddey RM (2012) Evidence from field nitrogen balance and ^{15}N natural abundance data for the contribution of biological N_2 fixation to Brazilian sugarcane varieties. Plant Soil 356:5–21

Viera-Vargas MS, Oliveira OC, Souto CM, Cadisch G, Urquiaga S, Boddey RM (1995) Use of different ^{15}N labelling techniques to quantify the contribution of biological N_2 fixation to legumes. Soil Biol Biochem 27(9):1185–1192

Wang L, Good S, Caylor K, Cernusak L (2012) Direct quantification of leaf transpiration isotopic composition. Agric For Meteorol 154–155:127–134

Witty JF, Renne RJ, Atkins CA (1988) ^{15}N addition methods for assessing N_2 fixation under field conditions. In: Summerfield RJ (ed) World crops: cool season food legumes. Kluwer Academic, Dordrecht, pp 716–730

Xing GX, Cao YC, Sun GQ (1997) Natural ^{15}N abundance in soils. In: Zhu ZL, Wen Q, Freney JR (eds) Nitrogen in soils of China. Kluwer Academic, London, pp 31–41

Yakir D, Sternberg L (2000) The use of stable isotopes to study ecosystem gas exchange. Oecologica 123(3):297–311

Zaman M, Zaman S, Adhinarayanan C, Nguyen ML, Nawaz S (2013b) Effects of urease and nitrification inhibitors on the efficient use of urea for pastoral systems. Soil Sci Plant Nutr 59:649–659

Zaman M, Saggar S, Stafford AD (2013a) Mitigation of ammonia losses from urea applied to a pastoral system: the effect of nBTPT and timing and amount of irrigation. N Z Grassl Assoc 75:121–126

Zaman M, Barbour MM, Turnbull MH, Kurepin LV (2014) Influence of fine particle suspension of urea and urease inhibitor on nitrogen and water use efficiency in grassland using nuclear techniques. In: Heng LK, Sakadeva K, Dercon G, Nguyen ML (eds) International Symposium on managing soils for food security and climate change adaption and mitigation. Food and Agriculture Organization of the United Nations, Rome, pp 29–32

www.ingramcontent.com/pod-product-compliance
Lightning Source LLC
LaVergne TN
LVHW082234200926
848484LV00008B/2354